One Health Perspective on Antimicrobial Resistance and Some Strategies for its Mitigation

One Health Perspective on
Antimicrobial Resistance and Some Strategies for its Mitigation

Editors

Hari Mohan Saxena
DRI-cum-Dean PGS
Bihar Animal Sciences University (BASU), Patna, India

Simeon Fogue Kouam
Professor and Natural Product Chemist
Department of Chemistry
Higher Teacher Training College
University of Yaounde I, Cameroon

A Publication of the Centre for Science & Technology of the Non-Aligned and Other Developing Countries (NAM S&T Centre)

ALLIED PUBLISHERS PVT. LTD.

New Delhi • Mumbai • Kolkata • Chennai • Bangalore • Hyderabad

ALLIED PUBLISHERS PRIVATE LIMITED

1/13-14 Asaf Ali Road, **New Delhi**–110002
Ph.: 011-23239001 • E-mail: delhi.books@alliedpublishers.com

17 Chittaranjan Avenue, **Kolkata**–700072
Ph.: 033-22129618 • E-mail: cal.books@alliedpublishers.com

15 J.N. Heredia Marg, Ballard Estate, **Mumbai**–400001
Ph.: 022-42126969 • E-mail: mumbai.books@alliedpublishers.com

No. 25/10, Commander-in-Chief Road, Ethiraj Lane (Next to Post Office)
Egmore, **Chennai**–600008
Ph.: 044-28223938 • E-mail: chennai.books@alliedpublishers.com

P.B. No. 9932, No. 15, 3rd Floor (Next to Vijaya Bank), 5th Cross, Gandhinagar, Karnataka,
Bangalore–560009, Ph.: 080-41530285 / 22386239
• E-mail: bngl.journals@alliedpublishers.com / apsabng@airtelmail.in

Sri Jayalakshmi Nilayam, No. 3-4-510, 3rd Floor (Above More Super Market)
Barkatpura, **Hyderabad**–500027
Ph.: 040-27551811, 040-27551812 • E-mail: hyd.books@alliedpublishers.com

Website: www.alliedpublishers.com

A Publication of the Centre for Science & Technology of the
Non-Aligned and Other Developing Countries (NAM S&T Centre), New Delhi

ISBN: 978-93-89934-16-8

Contributing Staff Editors: **Jasmeet Kaur**, Programme Officer & **Nidhi Utreja**, Research Associate
NAM S&T Centre, New Delhi

Foreword

Antimicrobial Resistance (AMR) occurs naturally over time when microorganisms (such as bacteria, fungi, viruses, and parasites) go through genetic changes when exposed to antimicrobial drugs. Microorganisms that develop antimicrobial resistance are sometimes referred to as "superbugs". As a result, the medicines become ineffective and infections persist in the body.

Antimicrobial resistant-microbes are found in people, animals, food and the environment (in water, soil and air). They can transmit between humans and animals; which includes from food of animal origin, and from a person to another. New resistance mechanisms are emerging and spreading globally, threatening the ability of anti-microbial agents to treat common infectious diseases, resulting in prolonged illness, disability, and death. Without effective anti-microbials for prevention and treatment of infections, even medical procedures have very high risk. Antimicrobial resistance is even putting the gains of the Millennium Development Goals at risk while endangering the achievement of the Sustainable Development Goals of 2030.

In order to discuss and deliberate on the current global, national and regional status of "Antimicrobial Resistance", the Centre for Science and Technology of the Non-Aligned and Other Developing Countries (NAM S&T Centre) had organised a three-day dynamic International Training Workshop on 'Antimicrobial Resistance and Strategies for its Mitigation' in collaboration with Guru Angad Dev Veterinary & Animal Sciences University (GADVASU), Ludhiana, during 10–13 July, 2018 in Ludhiana (Punjab), India with the objective to evolve strategies for mitigation of microbial resistance in humans and animals to antimicrobial agents and foster international, regional and sub-regional cooperation and collaboration in its effective implementation. It also aimed to focus the attention of the non-aligned and other developing on the global problem of ever increasing resistance of microorganisms to various antimicrobial agents.

As a follow up, the NAM S&T Centre has brought out a book titled "One Health Perspective on Antimicrobial Resistance and Some Strategies for its Mitigation" which comprises 15 papers contributed by the participants of the Ludhiana Workshop and other experts in this field. The volume has been edited by Dr. Hari Mohan Saxena, DRI-cum-Dean PGS, Bihar Animal Sciences University (BASU), Patna, India and Dr. Simeon Fogue Kouam, Professor and Natural Product Chemist, Department of Chemistry, Higher Teacher Training College, University of Yaounde I, Cameroon. The papers included in this book highlight the current status and the extent of antimicrobial resistance in humans and animals; focusing on evolving new and effective mitigation strategies for combating antimicrobial resistance in pathogenic microbes.

I congratulate the NAM S&T Centre for bringing out this valuable publication. It will be of great value to microbiologists, immunologists, virologists, biotechnologists, biochemists and other healthcare professionals working in the academic and research institutions, government departments and ministries and other relevant industries alike.

I wish the very best for all future endeavours of the NAM S&T Centre towards enhancing South-South S&T Cooperation amongst the NAM and other developing countries.

Prof Mahmoud M. Sakr
President
Academy of Scientific Research & Technology
ASRT

١٠١ شارع القصر العيني - القاهرة - مصر تليفون: ٢٧٩٢١٢٨٦ - ٢٧٩٢١٢٨٧ (٢٠٢+) فاكس: ٢٧٩٢١٢٧٠ (٢٠٢+) الرقم البريدي: ١١٥١٦

Introduction

Antimicrobial Resistance (AMR) is one of the most serious global public health concerns and profoundly threatening our ability to treat a wide range of infectious diseases caused by bacteria, parasites, viruses and fungi. It is occurring everywhere in the world and undermining many global advances in health and medicine. It is not only threatening the very core of modern medicine but also the sustainability of an effective public health response to the enduring threat from the ever-increasing range of infections. Systematic misuse and overuse of the antimicrobial drugs have put every nation at risk.

Addressing the rising threat of antimicrobial resistance requires a holistic and multi-sectoral approach referred to as **'One Health'** because antimicrobials used to treat various infectious diseases in animals may be the same or similar to those used for humans as well. Resistant bacteria arising in humans, animals or the environment may spread from one to the other, and from one country to another as AMR does not recognize geographic limitations and human-animal borders.

Alert to this crisis, the World Health Assembly held in May 2015 adopted a global action plan to tackle antimicrobial resistance, including antibiotic resistance; the most urgent drug resistance trend.

The global action plan ensures, for as long as possible, the continuity of successful treatment and prevention of infectious diseases with effective and safe medicines that are quality-assured, used in a responsible way, and accessible to all who need them. The plan outlines: to improve awareness and understanding of antimicrobial resistance, to strengthen knowledge through surveillance and research, to reduce the incidence of infection, to optimize the use of antimicrobial agents, develop the economic situation for sustainable investment that takes account of the needs of all countries, and increase investment in new R&D programmes on drug development, new medicines, diagnostic tools, vaccines and other related interventions.

In order to focus the attention of the *non-aligned and other developing countries* on this ever-increasing global problem of antimicrobial resistance, the *Centre for Science and Technology of the Non-Aligned and Other Developing Countries (NAM S&T Centre), New Delhi*, in partnership with the *Guru Angad Dev Veterinary & Animal Sciences University (GADVASU), Ludhiana* organised an International Training Workshop on '**Antimicrobial Resistance and Strategies for its Mitigation**' during 10–13 July, 2018 in Ludhiana (Punjab), India. The three-day Workshop brought together 53 researchers and scientists from 15 countries namely: Afghanistan, Egypt, Indonesia, Iran, Kenya, Malaysia, Mauritius, Myanmar, Nepal, Nigeria, Palestine, South Africa, Sri Lanka and Zambia and the host country India with an objective to identify the current status of antimicrobial resistance in NAM and other developing countries, specifically focusing to discuss antimicrobial resistance in animal and human pathogens; particularly the zoonotic infections, nosocomial and other chronic infections, reasons for vaccine failures and consequently evolve and implement strategies for its mitigation.

As a follow up of the Ludhiana Workshop, the present book titled "**One Health Perspective on Antimicrobial Resistance and Some Strategies for its Mitigation**" is being published by the NAM S&T Centre which includes 15 scientific papers contributed by the participants and other professionals and experts from seven developing countries namely: India, Indonesia, Mongolia, Nepal, Nigeria, South Africa and Sri Lanka.

I highly appreciate the time and efforts put in place by the Editors of this book, Dr. Hari Mohan Saxena, DRI-cum-Dean (PGS), Bihar Animal Sciences University (BASU), Patna, India and Dr. Simeon Fogue Kouam, Professor and Natural Product Chemist, Department of Chemistry, Higher Teacher Training College, University of Yaounde I, Cameroon, for the technical editing of the manuscripts.

I take this opportunity to express my gratitude to Prof. Mahmoud M. Sakr, President, The Academy of Scientific Research and Technology (ASRT), Cairo, Egypt for writing the "Foreword" for this book in spite of his very busy schedule.

I also acknowledge the interest and valuable efforts of the entire team of the NAM S&T Centre, and especially thank Mr. M. Bandyopadhyay, Senior Adviser for his guidance and supervision; and Ms. Jasmeet Kaur, Programme Officer; Ms. Nidhi Utreja, Research Associate; and Mr. Pankaj Buttan, Data Processing Manager for compiling and proof reading of the manuscripts, liaising with authors of the papers, designing the cover page, formatting and taking all the necessary steps in giving a shape to this book.

A special thanks to Prof. Arun P. Kulshreshtha, Former Director-General, NAM S&T Centre, for his interest and support in organising this important and valuable event.

I am also thankful to Mr. Sharad Gupta, Publishing Consultant and Mr. Jagdish Singh from the Publication Department, Allied Publishers, New Delhi, for their significant efforts in bringing out this valuable publication.

I am sure this book will be of great significance and immense use to all those associated with the multifaceted phenomenon of anti-microbial resistance.

(Amitava Bandopadhyay, *Ph.D.*)
Director General
NAM S&T Centre
New Delhi

Preface

Antimicrobial Resistance (AMR) is a major global public health issue. AMR is responsible for approximately 700,000 deaths each year, and is expected to cause one death every three seconds worldwide by 2050. The development of bacterial resistance to one or several antimicrobials is associated with the over-use of antimicrobials in human and veterinary medicine. Antimicrobials have played an important role in controlling bacterial infectious diseases in both humans and animals. In livestock, antimicrobials are used mainly for the treatment and prevention of diseases. The worldwide consumption of antimicrobials in animal-food production has been reported at around 57 million kg. Antimicrobials are not fully metabolized when administered to either humans or livestock. Upto 90% of many of the antibiotics used in livestock are excreted in urine or feces. Bacteria residing in the gastrointestinal tract may become resistant to these antibiotics and, once released into the environment, they may transfer Antimicrobial Resistance Genes (ARGs) to other bacteria including potential human pathogens. Furthermore, residual antibiotics may enter the environment through run-off from manure, where they may select for antimicrobial resistant bacteria. For almost every livestock-associated bacterial pathogen, resistance to at least one antimicrobial from each antimicrobial class has been reported. Antimicrobials, anti-microbial resistant bacteria and antimicrobial resistance genes are frequently detected in Sewage Treatment Plants (STPs) and as a result these can act as a potential hotspot for antibiotic resistance, where ARGs spread among bacteria via horizontal gene transfer. These biological pollutants are also released into the environment in STP effluent.

Antimicrobial Resistance is a global public health threat, and in the United States alone at least 23,000 people die each year due to resistant bacterial infections. It is also a 'One Health' issue because AMR emergence in bacteria from humans, animals, or the environment can impact the health of the others. As such, it is critical to identify and characterize emerging AMR threats in different parts of the world so that integrated control policies may be developed. It is essential that data from animal pathogens collected by veterinary diagnostic laboratories be incorporated into AMR surveillance activities as part of the One Health framework. These data, from bacterial pathogens of clinically ill animals, are an important addition to other surveillance programs that look at bacteria from healthy farm animals, foods and ill humans. Including veterinary pathogens in AMR surveillance will directly assist the veterinary profession treating our companion animals and will indirectly enhance our understanding of the epidemiology of AMR. The data from such studies can also be used to develop Antimicrobial Use (AMU) guidelines to educate veterinarians on the principles of good antimicrobial stewardship in their daily practice. Since the health of humans and animals are intricately linked, this data source is one of the critical components of One Health surveillance.

Medical doctors and veterinarians should emphasize the prudent use of antimicrobials. Antibiotics should be strictly used only when required. In addition to the conventional prevention strategies, special emphasis is required in the use of alternative prevention strategies, including vaccines, prebiotics, probiotics, and herbal drugs, to improve the production performance and health status of livestock. In beef production, antimicrobials are important to maintain or improve animal health and increase productivity. Although the development of AMR is a complex multifactorial process, use of potent broad-spectrum antimicrobials is a key factor for its development. The use of bacteriophage-based products, vaccines as well as other infection prevention and control approaches are promising alternatives to antimicrobials. One Health approaches to optimization of AMU have been suggested as measures for prolonging the therapeutic life of available antimicrobial drugs. A collective action towards promoting the prudent/judicious use of antimicrobials and search for alternative therapies is being advocated on a global scale.

Considering the global importance of the problem of ever increasing Antimicrobial Resistance, the Centre for Science and Technology of the Non-Aligned and Other Developing Countries (NAM S&T Centre). New Delhi and Guru Angad Dev Veterinary and Animal Sciences University (GADVASU), Ludhiana, took the initiative to organize an International Training Workshop on 'Antimicrobial Resistance and Strategies for its Mitigation' in July, 2018 at Ludhiana (Punjab), India. Medical doctors, veterinarians and scientists from 15 non-aligned and other developing countries participated in the

Workshop and presented their findings on the prevailing scenario of AMR in their regions and shared some of their experiences with various strategies for its mitigation. It was decided to make these presentations available globally for the benefit of other scientists and researchers. The present book is a collection of selected articles on the topic contributed by the participants. It is hoped that this compilation will be useful and foster international and interdisciplinary collaborations for further research and action to combat the menace of AMR among pathogens of humans and livestock.

Hari Mohan Saxena, *Ph.D.*

Simeon Fogue Kouam, *Ph.D.*

Contents

SECTION 1

Antimicrobial Resistance (AMR)

Antibiotic Resistance Profiling of Animal Origin *Escherichia coli* Isolates Associated with Various Clinical Infections

Sandeep Kumar Sharma[1,2*], Farah Naz Faridi[1], Vikas Galav[1,3] and Manish Agrawal[1,3]

[1]Veterinary Microbiology, Office of the RKVY Project Epidemiological Mapping of Antimicrobial Resistance (EMAMR)

[2]Department of Veterinary Microbiology and Biotechnology,

[3]Department of Veterinary Pathology

Post Graduate Institute of Veterinary Education and Research (PGIVER)

Rajasthan University of Veterinary and Animal Sciences (RAJUVAS), Jaipur (Rajasthan), India

E-mail: *drsharmask01@hotmail.com

ABSTRACT: The extensive and indiscriminate use of antimicrobial agents for prophylaxis and therapy management of ailments is responsible for the rapid emergence of resistance. Escherichia coli *is a well-established indicator bacterial pathogen for monitoring of antibiotic resistance and to analyze the underlying mechanisms of multidrug resistance. Therefore, the present study was designed to determine antibiotic resistance profile, minimum inhibitory concentration, ESBL phenotypes and MAR value of Escherichia coli isolates that were obtained from various clinical affections of animals. A total of 90 genetically confirmed animal origin* E. coli *isolates were screened against 61 antibiotics of almost all the available groups. All isolates were detected as multidrug resistant (MDR) isolates and 90.0% isolates were resistant to most of β-lactams, cephalosporins, carbapenem and penem group antibiotics such as penicillin, piperacillin, cefipime, cefotaxime, cefonicid, cefaclor, cefpodoxime, cefprozil, cephalothin, cefotaxime-clavulanate, ceftazidime-clavulanate imipenem and faropenem. The only cephalosporin antibiotic, cefmetazole was found to be 70.0% effective while 63.33% isolates were intermediate for chloramphenicol. Ninety percent of the isolates were sensitive towards fosfomycin, making it the most effective antibiotic. During ESBL screening, 55.55% isolates were detected as ESBL producers while remaining others were detected as non-producers. Further, MIC range was detected as > 256.0 mcg/ml, 0.19 to > 256.0 mcg/ml and 12.0 to > 256.0 mcg/ml for cefotaxime, gentamicin and ciprofloxacin respectively for studied isolates. The detected higher range of MIC indicates the excess use of routine antibiotics. In an effort for risk assessment of MDR isolates, MAR value was calculated. In the present study dog, cattle, poultry, sheep and goat isolates were detected with 0.72, 0.67, 0.66, 0.57 and 0.50 group MAR index respectively. The higher index is asserting the capability of an isolate as potential source for spread of MDR. This study further suggests genotypic characterization of isolates with regards to antibiotic resistant genes.*

Keywords: *Escherichia coli*, Animal Origin, Antibiotic Resistance, ESBL, MIC, MAR Index.

1. INTRODUCTION

Animals are the most important reservoir of pathogenic *Escherichia coli,* which is capable of causing several intestinal and extra-intestinal infections like diarrhea, urinary tract infection, meningitis, peritonitis, septicemia, and gram-negative bacterial pneumonia in animals besides being zoonotic also (Cooke *et al.*, 1971). This opportunistic bacterium is an integral and predominant part of normal resident bacterial flora in the intestine of humans and animals used for food production. It plays a very important role as an indicator of antimicrobial selective pressure (Harada and Tetsuo, 2010). The significant use of antibiotics for treatment of veterinary diseases, prophylaxis and growth promotion has led to the development of reservoir of multi drug resistant *E. coli* in animals. The World Health Organization (WHO) has ranked antimicrobial resistance as one of the three greatest threats to human health. The exploding emergence of multidrug resistance in *E. coli* globally is of grave concern as multi-drug resistant organisms (MDROs) further complicate the diseased condition and results in further decreased effectiveness of standard treatments and increased healthcare costs (Swami *et al.*, 2012). Bacteria from the animal reservoir carrying resistance to antimicrobial agents regarded as highly or critically important to human therapy like aminoglycosides, fluoroquinolones, and third- and fourth-generation cephalosporins are of special concern as they are responsible for the spread of the bacterial resistance genes from animals to humans (Anette and Heuer, 2009). Keeping the role of *E. coli* with regards to antibiotic resistance and healthcare concern of humans and animals in mind, the drug resistant *E. coli* phenotypes should be identified and their underlying genetic mechanisms governing resistance should be investigated (Aslam *et al.*, 2009). Therefore, the following study was undertaken to (i) study the resistance profile of *E. coli* obtained from clinical samples (ii) to detect ESBL phenotypes and MIC for the resistant isolates (iii) to calculate MAR value index for the animal origin *E. coli* isolates.

2. MATERIALS AND METHODS

2.1 Collection of Samples

A total of 120 clinical samples of animal origin in and around Jaipur, Rajasthan, India were preceded for selective isolation of *Escherichia coli.* Clinical samples of cattle, sheep, goat, dog and poultry which belonged to diarrheic feces, pneumonic nasal discharge, vaginal/uterine clinical discharge, ear exudates, mastitic milk, and pus exudates from wound infections were enriched in nutrient broth overnight at 37°C for 18–24 hours.

2.2 Isolation and Identification of Bacteria

The overnight broth inoculated samples were streaked on MacConkey agar and red pin point lactose fermenting colonies were further taken on nutrient agar for obtaining pure cultures. The pure cultures were further characterized by primary and secondary biochemical test such as Gram's staining, oxidase test, catalase test. Isolates were also checked for metallic sheen on EMB agar and IMViC pattern was also studied. The genomic DNA was isolated from pure colonies using snap chill method (Alexopoulou *et al.*, 2006). Briefly, pure bacterial colony was lysed in millipore water by heating in boiling water bath for 10 minutes followed by snap chilling at 4°C or below for 2 minutes. The suspension was centrifuged to obtain genomic DNA in supernatant. *E. coli* was confirmed through PCR using supernatant as a DNA template and specific primers as described by Khaled A. *et al.*, 2010.

2.3 Antibiotic Sensitivity Test

The antibiotic resistance pattern was determined for all the genotypically confirmed isolates. The isolates were screened for 61 antibiotics recommended by CLSI 2017, belonging to all different groups available. The disks were procured from manufacturers (Himedia Pvt. Ltd., Mumbai) as per the recommended guidelines of CLSI (2017). The standard agar disc diffusion method described by Bauer-Kirby (Bauer *et al.*, 1966) using Mueller Hinton agar was followed to determine resistance pattern.

2.4 Extended-Spectrum β-lactamase (ESBL) Production

The resistant (genotypically confirmed) isolates were further screened for the production of β-lactamase by using Hichrome ESBL agar base (Himedia Pvt. Ltd., Mumbai) and ESBL selective supplement. The media was prepared according to the instructions given by the manufacturer. Pure colonies of the resistant *E. coli* were streaked on the ESBL agar base and incubated overnight as per the method described by the manufacturer.

2.5 Minimum Inhibitory Concentration (MIC) Detection

The five most resistant isolates were considered for MIC detection (in mcg/ml) for Cefotaxime, Gentamicin and Ciprofloxacin. The test was performed using commercially available Ezy MIC™ detection strips (Himedia Pvt. Ltd., Mumbai) having a pre-defined quantitative gradient. The appropriate strip was placed on MHA plates which were already swabbed with *E. coli* culture and protocol was followed as per the method described by the manufacturer. The results were read where the ellipse intersected the MIC scale on the strip after incubation. Since, Ezy MIC™ strip has continuous gradient, MIC values "in-between" two fold dilutions was obtained. If the ellipse intersected the strip in between 2 dilutions, the MIC value nearest to the zone was recorded.

2.6 Multiple Antibiotic Resistance (MAR) Index Value

In an effort for risk assessment of MDR isolates, this index was calculated as per the method given by Krumperman (1983). MAR index of individual isolate is identified as,

$$\text{MAR index} = a/b \qquad \dots (1)$$

Where a is the number of antibiotics to which the isolate was resistant, b is the number of antibiotics to which the isolate was exposed,

$$\text{Group MAR index} = a/(b \times c) \qquad \dots (2)$$

Where a is the aggregate antibiotic resistance score of all isolates in the group, b is the number of antibiotics to which the isolate was exposed, c is the total number of isolates in the group.

3. RESULTS

Of the total 120 samples of various clinical sources, a total of ninety *Escherichia coli* isolates were obtained. All ninety isolates were confirmed by phenotypic tests and genotypic tests by species specific PCR. All isolates were detected as red pink lactose fermenting colonies on MacConkey agar, metallic sheen was observed on EMB agar, with ++– – IMViC pattern and a 662 bp amplicon was observed with species specific PCR (Figure 1).

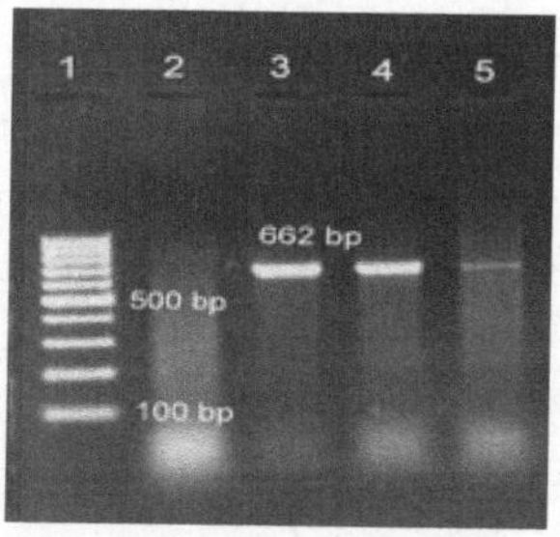

Figure 1: Lane 1–100 bp ladder, Lane 3 to 5-*Escherichia coli* (662 bp amplicon)

3.1 Antibiotic Sensitivity Pattern

All ninety isolates were screened with the 61 recommended antibiotics and results were recorded as sensitive, intermediate or resistant according to the guidelines of Clinical and Laboratory Standards Institute (CLSI, 2017) including interpretations. All isolates were detected as multidrug resistant (MDR) isolates. Out of the total isolates, 90.0% were resistant to most of β-lactams, cephalosporins, carbapenem and penem group of antibiotics such as penicillin, piperacillin, cefipime, cefotaxime, cefonicid, cefaclor, cefpodoxime, cefprozil, cephalothin, cefotaxime-clavulanate, ceftazidime-clavulanate, imipenem and faropenem. Cefmetazole was the only cephalosporin antibiotic to which the 70% isolates were sensitive while 63.33% isolates were intermediate for chloramphenicol. Out of the total 61 antibiotics, fosfomycin was detected as the most effective antibiotic, to which 90.0% of the isolates were recorded sensitive. Detailed results are described in Table 1.

Table 1: Antibiotic Sensitivity Pattern of *Escherichia coli* Isolates

S. No.	*Group of Antibiotics*	*Name of Antibiotics (Conc.- μg)*	*Percentage of Isolates*		
			Resistant	*Intermediate*	*Sensitive*
1.	Penicillins	Ampicillin (10)	88.88	5.55	5.55
		Piperacillin (100)	93.33	6.66	0.0
		Mecillinam (10)	44.44	11.11	44.44
2.	β-Lactam/ β-Lactamase Inhibitor Combinations	Amoxycillin/ Clavulanate (20/10)	36.66	38.88	24.44
		Ampicillin/ Sulbactam (10/10)	30.0	44.44	25.55
		Ticarcillin/ Clavulanate (75/10)	63.33	25.55	11.11
		Piperacillin/ tazobactam (100/10)	84.44	12.22	3.33
3.	CEPHEMS (Including Cephalosporins I, II, III and IV generation	Cefazolin (30)	65.55	10.0	24.44
		Cefipime (30)	96.66	0.0	3.33
		Cefotaxime (30)	98.88	0.0	1.11
		Ceftriaxone (30)	65.55	11.11	23.33
		Cefotetan (30)	57.77	38.88	3.33
		Cefuroxime (30)	71.11	24.44	4.44
		Cefoxitin (30)	46.66	28.88	24.44
		Ceftazidime (30)	83.33	15.55	1.11
		Cefamandole (30)	57.77	21.11	21.11
		Cefmetazole (30)	18.88	11.11	70.0
		Cefonicid (30)	98.88	0.0	1.11
		Cefoperazone (75)	64.44	26.66	8.88
		Ceftizoxime (30)	64.44	23.33	12.22
		Moxalactam (30)	47.77	50.0	2.22
		Cefaclor (30)	92.22	5.55	2.22
		Cefedinir (5)	86.66	8.88	4.44
		Cefixime (5)	72.22	15.55	12.22
		Cefpodoxime (10)	96.66	1.11	2.22
		Cefprozil (30)	93.33	5.55	1.11
		Cephalothin (30)	95.55	2.22	2.22
		Cephalexin (CN)	66.66	0.0	33.33
4.	CEPHEMS/ β-Lactamase Inhibitor Combinations	Cefotaxime/Clavul anate (20/10)	97.77	0.0	2.22
		Ceftazidime/ Clavulanate (20/10)	98.88	0.0	1.11

S. No.	*Group of Antibiotics*	*Name of Antibiotics (Conc.- µg)*	*Percentage of Isolates*		
			Resistant	*Intermediate*	*Sensitive*
5.	Monobactams	Aztreonam (30)	54.44	23.33	22.22
6.	Carbapenems	Imipenem (10)	93.33	0.0	17.77
		Meropenem (10)	80.0	10.0	10.0
		Doripenem (10)	41.11	26.66	32.22
		Ertapenem (10)	40.0	26.66	33.33
7.	Penem	Faropenem (FAR)	93.33	0.0	5.55
8.	Amino-glycosides	Gentamicin (10)	48.88	34.44	18.88
		Tobramycin (10)	30.0	34.44	35.55
		Amikacin (30)	71.11	23.33	5.55
		Kanamycin (30)	44.44	43.33	12.22
		Streptomycin (10)	41.11	38.88	20.0
		Netillin/ Netilmicin (30)	40.0	32.22	27.77
9.	Macrolides	Azithromycin (15)	43.33	5.55	41.11
10.	Tetracyclines	Tetracycline (30)	61.11	12.22	26.66
		Minocycline (30)	55.55	37.77	6.66
		Doxycycline Hydrochloride (30)	51.11	15.55	33.33
11.	Quinolones and Fluoro-quinolones	Ciprofloxacin (5)	54.44	18.88	26.66
		Levofloxacin (5)	55.55	14.44	30.0
		Cinoxacin (100)	66.66	6.66	26.66
		Enoxacin (10)	47.77	10.0	42.22
		Gatifloxacin (5)	56.66	14.44	28.88
		Gemifloxacin (5)	65.55	13.33	21.11
		Lomeofloxacin (10)	87.77	6.66	5.55
		Norfloxacin (10)	45.55	18.88	36.66
		Ofloxacin (5)	65.55	15.55	26.66
		Nalidixic Acid (30)	68.88	20.0	11.11
12.	Folate Pathway Inhibitors	Trimethoprim (5)	47.77	7.77	44.44
		Sulfisoxazole (SF)	48.88	14.44	36.66
13.	Phenicols	Chloramphenicol (30)	24.44	12.22	63.33
14.	Fosfomycins	Fosfomycin (200)	3.33	5.55	91.11
15.	Nitrofurans	Nitrofurantoin (300)	54.44	34.44	11.11

3.2 Extended-Spectrum β-lactamase (ESBL) Production

During ESBL screening, 55.55% isolates (50 out of 90) were detected as ESBL producers while remaining others were recorded as non ESBL producers (Figure 2). Pink to purple colored luxuriant growth of organisms was considered as ESBL producing *E. coli.*

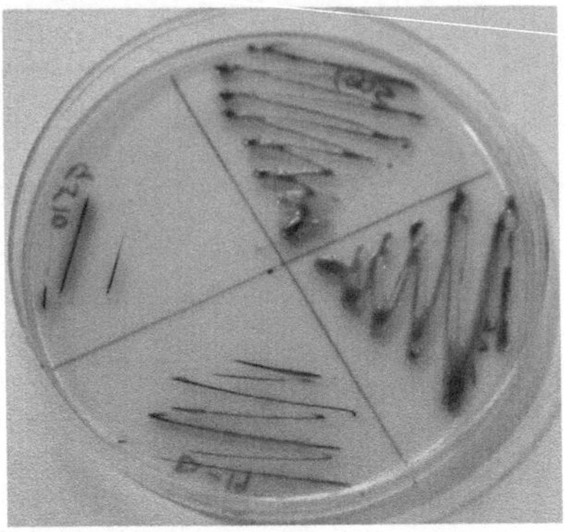

Figure 2: A Plate Showing ESBL Producing *Escherichia coli* Isolates

3.3 Minimum Inhibitory Concentration (MIC)

For quantitative determination of susceptibility some selected isolates were screened for MIC (Figure 3). The isolates were checked for three antibiotics and MIC range was detected as > 256.0 mcg/ml, 0.19 to > 256.0 mcg/ml and 12.0 to > 256.0 mcg/ml for cefotaxime, gentamicin and ciprofloxacin respectively. The detailed MIC results are described in Table 2.

Table 2: Minimum Inhibitory Concentration (MIC) Determination for *Escherichia coli* Isolates

Isolate ID	*Name of Antibiotic (conc./ml)*		
	Cefotaxime (mcg/ml)	*Gentamicin (mcg/ml)*	*Ciprofloxacin (mcg/ml)*
Dog 09	> 256.0	4.0	> 256.0
Sheep (a)	> 256.0	> 256.0	> 256.0
Sheep5 (b)	> 256.0	0.38	> 256.0
Goat 41	> 256.0	0.19	> 256.0
Sheep 157	> 256.0	1.5	12.0
Overall Range	> 256.0	0.19 to >256.0	12.0 to > 256.0

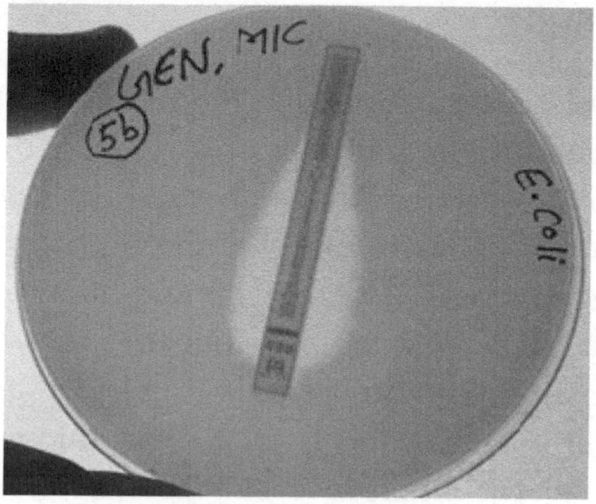

Figure 3: Minimum Inhibitory Concentration (MIC) of *Escherichia coli* for Gentamicin

3.4 Multiple Antibiotic Resistance (MAR) Index

In the present study, isolates were subjected to MAR evaluation to find out their capability to spread antibiotic resistance among other isolates. Hence, individual MAR was detected and it was found that all (100%) studied isolates had more than 0.2 MAR index (0.30 to 0.98 value) while none of the isolate was present with MAR value less than 0.2. Further, the isolates were evaluated for group MAR and dog, cattle, poultry, sheep and goat isolates were detected with 0.72, 0.67, 0.66, 0.57 and 0.50 group MAR index respectively, as described in Table 3. This high value of individual as well as group MAR indicates that the isolates have more capacity to spread resistance and can be the potential hazard for the society.

Table 3: Group MAR Index Value of *Escherichia coli* Isolates

Source of Isolates	*Number of Isolates*	*Aggregate Resistance Antibiotic Score*	*Group MAR Index*	*Significance*
Dog	23	1015	0.72	Greater than 0.2 MAR index of group indicates that several antibiotics have been used in that group and the bacteria have a high potential risk for the spread of multi drug resistance.
Cattle	12	491	0.67	
Poultry	17	694	0.66	
Sheep	26	913	0.57	
Goat	12	366	0.50	

4. DISCUSSION

Due to close involvement of animals and humans, the antibiotic resistant bacteria of zoonotic importance pose a severe threat to mankind. There is no denying the fact that the extensive use of antibiotics is accepted as a major selective force for the high prevalence of multi drug resistant bacteria both in veterinary and human medicine. There are reports on the data from various countries that had also reported higher resistance rates in poultry as well as in other animals (Ilias *et al.*, 2014). Therapeutic management of *E. coli* infections has become problematic due to the emergence of resistance to most of first-line antimicrobial drugs. The current work revealed prevalence of multi drug resistant *E. coli* isolated from clinical animal settings and more than 90% of isolates were found resistant to most of the Cephalosporin antibiotics. Resistance to Extended Spectrum Cephalosporins (ESCs) has been reported worldwide in *E. coli* isolated from human and other animal sources (Laura *et al.*, 2005, Xian *et al.*, 2007).

The high resistance towards cephalosporin has also been found in clinical samples from dogs in America with resistance frequency ranging from 54–70% (Kevin *et al.*, 2015). This does not come as a surprise because these drugs have been in use for the longest time both in human and veterinary medicine. Over the few decades, resistance to cephalosporins in *E. coli* has increased mainly due to the spread of extended-spectrum β-lactamases (ESBL) as ESBL genes are the most important mediators of cephalosporin resistance in *E. coli* and are relevant in terms of emerging resistance properties (Philippon *et al.*, 1994). The detection of ESBL producing strains has become significantly important for controlling infections and epidemiological surveillance. *Escherichia coli* along with other *Klebsiella* spp. are most common ESBL producing pathogens. The result of this study demonstrated that *E. coli* harboring ESBL enzymes are multi-drug resistant showing resistance to majority of the antibiotics tested and thus can pose a severe challenge to the public health. Many surveillance reports have been published reflecting the increased role of animals as ESBL reservoir. Production of β-lactamase in clinical strains of *E. coli* isolated from animals has also been reported in various (Carattoli *et al.*, 2005, Mulvey *et al.*, 2005, O'Keefe *et al.*, 2010). ESBL-producing *E. coli* has also been reported from the reproductive and GI tract of nine broiler breeding roosters in Algeria (Mezhoud *et al.*, 2015), in sheep, chicken, cattle, horse, rabbit and dromedaries from Tunisia (Ben Sallem *et al.*, 2012), in cattle from Egypt (Braun *et al.*, 2016). This study forms a premise that the ESBL production is one of the major resistance governing mechanism in *E. coli* isolated from veterinary clinical settings. The present study is in line of agreement with the already published studies that directly correlated the increased use of antibiotics with the emergence of resistance in bacteria (David *et al.*, 2002, Stuart and Bonnie, 2004). Fosfomycin was the only antibiotic that was recorded most effective against *E. coli* since its use is not reported as a regular antibiotic for veterinary therapeutics. The higher range of MIC detected indicates excess use of routine antibiotics in treatment of animals. Significant increases in MIC reading of Ciprofloxacin in *E. coli* were also reported by Wang *et al.* (2015). More than 0.2 individual and group MAR value implies that strains of such bacteria originate from an environment where it has been pre disposed to several antibiotics. All isolates group studied carry a high risk of spread of potential multi drug resistance in the environment as well as among other bacterial pathogens (Krumperman, 1983). The higher individual and group MAR value obtained indicated the risk of an isolate as a potential source of spread of multi drug resistance.

5. CONCLUSIONS

The inappropriate use of antibiotics is responsible for emergence, prevalence, and dissemination of multi drug resistant pathogenic *Escherichia coli*, which results in increased healthcare costs, duration, morbidity and mortality. This study provides epidemiological evidence that majority of the *E. coli* isolates studied above are at high risk of multi drug resistance as indicated by antibiotic sensitivity tests and ESBL production. Fosfomycin came out to be the most effective antibiotic. Further, over exploitation of antibiotics and their inadequate selection may lead to complete resistant *E. coli* and would make the treatment of its infections more difficult. The study further suggests the genotyping of isolates with regard to antibiotic resistance genes. The valuable information of resistance pattern, MIC and ESBL phenotypes of *E. coli* can be of immense significance in formulation of database and strategy to curb the menace of widespread antibacterial resistance.

ACKNOWLEDGEMENTS

This study was conducted under the financial support of Rashtriya Krishi Vikas Yojana (RKVY) project "Epidemiological Mapping of Antimicrobial Resistance (EMAMR) and authors would like to acknowledge the support and time to time encouragement provided by Dean, PGIVER (RAJUVAS), Jaipur (Rajasthan), India.

REFERENCES

[1] Alexopoulou, K., Foka, A., Petinaki, E., Jelastopulu, E., Dimitracopoulos, G. and Spiliopoulou, I. (2006). Comparison of two commercial methods with PCR restriction fragment length polymorphism of the *tuf* gene in the identification of coagulase-negative staphylococci. *Lett Appl Microbiol,* 43 (4): 450–454.

[2] Aslam, M., Diarra, M.S., Service, C. and Rempel, H. (2009). Antimicrobial resistance genes in *Escherichia coli* isolates recovered from a commercial beef processing plant. *J. Food Prot,* 72(5): 1089–1093.

[3] Bauer, A.W., Kirby, W.M.M., Sherris, J.C. and Turck, M. (1966). Antibiotic susceptibility testing by a standardized single disk method. *Am J Clin Path,* 45: 493–496.

[4] Ben Sallem, R., Ben Slama, K., Sáenz, Y., Rojo-Bezares, B., Estepa, V., Jouini, A., Gharsa, H., Klibi, N., Boudabous, A. and Torres, C. (2012). Prevalence and characterization of extended-spectrum beta-lactamase (ESBL)–and CMY-2–producing *Escherichia coli* isolates from healthy food-producing animals in Tunisia. *Foodborne Pathog Dis,* 9(12): 1137–1142.

[5] Braun, S.D., Ahmed, M.F., El-Adawy, H., Hotzel, H., Engelmann, I., Weib, D., Monecke, S. and Ehricht, R. (2016). Surveillance of extended-spectrum beta-lactamase-producing *Escherichia coli* in dairy cattle farms in the Nile Delta, *Egypt. Front microbial,* 1020.

[6] Brinas, L., Moreno, M.A., Teshager, T., Sáenz, Y., Porrero, M.C., Domínguez, L. and Torres, C. (2005). Monitoring and characterization of extended-spectrum β-lactamases in *Escherichia coli* strains from healthy and sick animals in Spain in 2003. *Antimicrob Agents Chemother,* 49(3):1262–1264.

[7] Carattoli, A., Lovari, S., Franco, A., Cordaro, G., Di Matteo, P. and Battisti, A. (2005). Extended-spectrum β-lactamases in *Escherichia coli* isolated from dogs and cats in Rome, Italy, from 2001 to 2003. *Antimicrob Agents Chemother,* 49(2): 833–835.

[8] Chantziaras, I., Boyen, F., Callens, B. and Dewulf, J. (2014). Correlation between veterinary antimicrobial use and antimicrobial resistance in food-producing animals: a report on seven countries. *J. Antimicrob Chemother,* 69(3): 827–834.

[9] CLSI (2017). Performance Standards for Antimicrobial Susceptibility Testing. 27th ed. CLSI supplement M100. Wayne, PA: Clinical and Laboratory Standards Institute.

[10] Cooke, E.M., AlwenaL, B., Shooter, R.A. and O'Farrell, S.M. (1971). Antibiotic sensitivity of *Escherichia coli* isolated from animals, food, hospital patients, and normal people. *Lancet,* 298 (7714): 8–10.

[11] Cummings, K.J., Aprea, V.A. and Altier, C. (2015). Antimicrobial resistance trends among canine *Escherichia coli* isolates obtained from clinical samples in the northeastern USA, 2004–2011. *Can Vet J.*, 56(4): 393–398.

[12] Hammerum, A.M. and Heuer, O.E. (2009). Human health hazards from antimicrobial resistant *Escherichia coli* of animal origin. *Clin Infect Dis,* 48 (7): 916–921.

[13] Harada, K. and Tetsuo, Asai, T. (2010). Role of Antimicrobial Selective Pressure and Secondary Factors on Antimicrobial Resistance Prevalence in *Escherichia coli* from Food-Producing Animals in Japan. *J. Biomed Biotechnol,* 2010:180682.

[14] Khaled A, Abd El-Razik, Khaled A, Abdelrahman, Youssef F, Ahmed., Alaa, Mohamed Gomaa and Hazem, A. Eldebaky (2010). Direct Identification of Major Pathogens of the Bubaline Subclinical Mastitis in Egypt using PCR. *J Am Sci.,* 6(10): 652–66.

[15] Krumperman, P.H. (1983). Multiple antibiotic resistance indexing of *Escherichia coli* to identify high-risk sources of fecal contamination of foods. *Appl Environ Microbiol,* 46(1): 165–170.

[16] Levy, S.B. and Marshall, B. (2004). Antibacterial resistance worldwide: causes, challenges and responses. *Nat Med,* 10 (12s). 122–129.

[17] Li, X.Z., Mehrotra, M., Ghimire, S. and Adewoye, L. (2007). β-lactam resistance and β-lactamases in bacteria of animal origin. *Vet Microbiol,* 121:197–214.

[18] Mezhoud, H., Boyen, F., Touazi, L.H., Garmyn, A., Moula, N., Smet, A., Haesbrouck, F., Martel, A., Iguerouada, M. and Touati, A. (2015). Extended spectrum β-lactamase producing *Escherichia coli* in broiler breeding roosters: presence in the reproductive tract and effect on sperm motility. *Anim Reprod Sci,* 159: 205–211.

[19] Mulvey, M.R., Bryce, E., Boyd, D.A., Ofner-Agostini, M., Land, A.M., Simor, A.E. and Paton, S. (2005). Molecular characterization of cefoxitin-resistant Escherichia coli from Canadian hospitals. *Antimicrob Agents Chemother,* 49: (1), 358–365.

[20] O'Keefe, A., Hutton, T.A., Schifferli, D.M. and Rankin, S.C. (2010). First detection of CTX-M and SHV extended-spectrum beta-lactamases in Escherichia coli urinary tract isolates from dogs and cats in the United States. Antimicrob Agents Chemother, 54(8):3489–92.

[21] Philippon, A., Arlet, G. and Lagrange, P.H. (1994). Origin and impact of plasmid-mediated extended-spectrum β-lactamases. *Eur J Clin Microbiol Infect Dis,* 13(1):17–29.

[22] Smith, D.L., Harris, A.D., Johnson, J.A., Silbergeld, E.K. and Morris, J.G. Jr. (2002). Animal antibiotic use has an early but important impact on the

emergence of antibiotic resistance in human commensal bacteria. *Proc Natl Acad Sci USA*, 99: 96434–6439.

[23] Swami, S.K., Liesinger, J.T., Shah, N., Baddour, L.M. and Banerjee, R. (2012). Incidence of antibiotic-resistant *Escherichia coli* bacteriuria according to age and location of onset: a population-based study from Olmsted County, *Minnesota. Mayo Clin Proc,* 87: 753–9.

[24] Wang, J.T., Chang, S.C., Chang, F.Y., Fung, C.P., Chuang, Y.C., Chen, Y.S., Shiau, Y.R., Tan, M.C., Wang, H.Y., Lai, J.F., Huang, I.W. and Yang Lauderdale, T.L. (2015). Antimicrobial Non-Susceptibility of *Escherichia coli* from Outpatients and Patients Visiting Emergency Rooms in Taiwan. PLoS one, 10 (12): 1–17. e0144103.

[25] WHO (2014). Antimicrobial resistance: global report on surveillance. World Health Organization, Geneva, Switzerland. http://www.who.int/drugresistance/documents/surveillancereport/en

Bacterial Resistance to Disinfectants: A Review of Biosecurity, Resistance Mechanisms and the Use of Sequencing Technology

Samantha McCarlie[1], Wouter van der Westhuizen[2], Jeffrey Newman[3], Julius Helmuth[4], Charlotte Boucher[5] and Robert Bragg[6*]

[1,2,4–6]Department of Microbial, Biochemical and Food Biotechnology, University of the Free State, P.O. Box 339, Bloemfontein 9300, South Africa

[2]Lycoming College, 700 College Place, Williamsport, Pennsylvania, PA 17701, USA

E-mail: *braggrr@ufs.ac.za

ABSTRACT: Antibiotic resistance is becoming one of the biggest life-threatening challenges of our time. Bacterial infections present in hospitals and agriculture are becoming unresponsive to many of the antibiotics currently in use, marking the start of a post-antibiotic era. One of the other lines of defence against bacterial growth and infections are disinfectants and improved biosecurity. This review looks at the emergence of disinfectant resistance as well as the mechanisms identified thus far. Currently, the transmembrane efflux pump is the most wide-spread mechanism of resistance to disinfectants, lowering the intracellular concentration of antimicrobial agents. This resistance mechanism is effective against both antibiotics and disinfectants in some multi-resistant strains, leading to cross-resistance. Resistance to disinfectants could prove difficult to prevent. However, resistance to disinfectants could be slowed down until other options for control of bacterial diseases can be developed and applied on an industrial scale. New sequencing technologies have vastly improved identification and the study of disinfectant resistance mechanisms. This is mainly due to the identification of resistance genes, their mechanisms and bacterial characterisation.

Keywords: Disinfectant Resistance, Biosecurity, Resistance Mechanisms, Cross-resistance, QAC Resistance Genes, Next Generation Sequencing.

1. INTRODUCTION

Bacterial resistance to antibiotics has become a significant issue. The resistance to antibiotics is disseminating rapidly due to the lack of proper biosecurity measures from the hospital environment to the food and agricultural industry (Hawkey and Jones, 2009, Bragg *et al.*, 2014, Spellberg and Gilbert, 2014, Ventola, 2015). Currently, the best viable protection we have left against bacteria is disinfectants (McDonnell and Russell 1999), however this makes it only more troubling that disinfectant resistance is also emerging at an alarming speed (Templeton *et al.*, 2009). In order to combat emerging resistance to disinfectants, we need to understand the mechanisms of resistance. Once the mechanisms are identified, possible solutions can be discussed (Khan *et al.*, 2016). If biosecurity is implemented properly, it can slow down the rate of developing resistance to disinfectants (Gross, 2013). As widespread resistance to disinfectants is highly likely, it is imperative that the other viable options for the control of bacterial growth need to be researched. The purpose of this paper is to consider bacterial resistance, particularly to disinfectants, to discuss the mechanisms of resistance, as well as sequencing technology to detect and monitor bacterial resistance to disinfectants.

2. THE SIGNIFICANCE OF BACTERIAL RESISTANCE TO ANTIBIOTICS AND BIOSECURITY

Antibiotics are used worldwide in many environments such as hospitals, agriculture, dentistry and veterinary practices. Antibiotics and disinfectants play an integral role in infection control and quality standards (Rutala, 1995, McDonnell and Russell, 1999). The average human life expectancy before antibiotics, in the early 20th Century, was 47 years in the United States of America, whereas currently it is 78.8 years (CDC 2016). This is largely due to the use of antibiotics to treat bacterial diseases such as cholera and typhoid fever which were devastating at the time (Adedeji, 2016). The rapid emergence of bacterial resistance to antibiotics is a significant threat to industry and life as we know it. The term 'resistant' can be used to describe an organism that is insusceptible to a concentration of disinfectant used in practice, or that is not inactivated or inhibited by a concentration that would otherwise inactivate the majority of strains of that organism (Russell 1987). Resistance to antibiotics is nothing new as bacteria such as *Streptomyces griseus* and yeasts like *Penicillium notatum* have been producing the antibiotics *Streptomycin* and *Penicillin*, respectively, for millions of years (Zaffiri *et al.*, 2012). In response to this, bacteria in the affected population develop the ability of resistance to these antibiotics. The use of any antimicrobial including antibiotics creates selective pressure allowing for the development of resistance in the affected population (Canton *et al.*, 2015).

The modern era of antibiotics started in 1928 when Sir Alexander Fleming discovered penicillin (Gould and Bal, 2013, Gross, 2013, Sengupta *et al.*, 2013). Thereafter more antibiotics and derivatives were developed, such as additional β-lactams and vancomycin. However, within 8 years of commercial use of these antibiotics, resistance was detected and these antibiotics no longer worked in some cases (Sengupta *et al.*, 2013, Spellberg and Gilbert 2014). The prevalence of antibiotic resistance is not limited to the medical environment. The agricultural industry uses antibiotics extensively on livestock, often irresponsibly which leads to accelerated rate of the emergence of resistance (Bragg *et al.*, 2014). Once antibiotic resistance has emerged, it easily spreads over vast areas in short amounts of time (Hawkey and Jones 2009, Davis *et al.*, 2015). Resistance genes such as the blaCTX-M family of ***Extended-Spectrum β-Lactamase (ESBL)*** genes have spread globally within 30 years in the dairy industry (Canton *et al.*, 2012, D'Andrea *et al.*, 2013). Antibiotic resistance like this is never isolated, but finds a way to weave into almost every industry. The presence of resistant strains reported in dairy cows may not be isolated to the dairy industry only. A study has shown that these resistance genes may be able to find their way into the food industry through the manure used to fertilise crop fields (Udikovic-Kolic *et al.*, 2014).

Currently there are little to no novel antibiotics produced as the pharmaceutical industry has very little financial incentive for their production (Allen *et al.*, 2014). Alternatives to antibiotics are numerous. Vaccines work to prevent both viral and bacterial infections by stimulating specific immune protection. Yet, there is limited cross-protection from pathogenic strains and it is costly to administer to an animal or human population (van Panhuis *et al.*, 2013). The mode of action of immune response in both animals and humans makes it very difficult to consistently obtain 100% protection with any bacterial vaccine. New generation vaccines and reverse vaccinology could improve the current levels of protection offered by first generation vaccines. However, many of these new generation vaccines are still in the developmental process. Alternatively, phage therapy could be applied for infectious disease treatment as they are target-specific and can be synergistic with antibiotics (Allen *et al.,* 2014). In addition, multiple bacteriophages can be used to hinder the development of resistance and studies have shown the efficacy of topical applications (Brussow 2007). However, there is potential for resistance development and the high levels of specificity of bacteriophages hampers administration against multiple subspecies (Johnson *et al.,* 2008, Chan *et al.,* 2013). Alternative treatment options such as bacteriocins, antimicrobial peptides and using competitive bacteria are possible but all of these methods are still in development, with their own safety concerns (Fox and Blommel 2009, Cotter *et al.,* 2013, Allen *et al.,* 2014). The 'post-antibiotic era' is upon us, with the only other viable option to be used on a large scale for industry being disinfectants. At the moment, good biosecurity and disinfection may be our only hope in a post-antibiotic era (McDonnell and Russell 1999).

3. RESISTANCE TO DISINFECTANTS AND THE IMPACT OF LACK OF PROPER BIOSECURITY MEASURES

The worldwide emergence of resistance is accelerated by the abuse and improper use of disinfectants (Bragg *et al.,* 2014). This provides the opportunity for the acquisition of genetic elements that confer resistance to other previously non-resistant strains (Hawkey and Jones 2009, Sengupta *et al.,* 2013, Bragg *et al.,* 2014). A lack of understanding of the principles behind the mechanism of disinfection and of biosecurity leads to inadequate use. This allows for disinfectant use at sub-MIC levels (Ventola, 2015). Over the time, this results in the selection of bacteria, able to withstand a given concentration of the disinfectant therefore, increasing the prevalence of resistant strains. A higher presence of these resistant strains in the population leads to contamination and mortalities (Spellberg and Gilbert, 2014).

The increased demand for less preservatives, less processed food and more ready-to-eat food; coupled with a lack of good biosecurity will inevitably lead to outbreaks of food borne disease. The South African outbreak of listeriosis in 2017 (caused by *Listeria monocytogenes*) resulted in a total of 674 confirmed cases and 183 reported deaths, although these numbers are assumed to be much greater in reality (WHO 2018). The source of the outbreak was traced back to a ready-to-eat, processed meat product called '***Polony***'. The food processing company and its associated retailers distribute this product throughout South Africa and to 15 other African countries. The economic implications of extensive recalls, widespread public panic and a loss of consumer confidence can be crippling for any company (Piet *et al.*, 2016). The risks by far outweigh the advantages of implementing good bio-security in any industry.

Biosecurity is the least understood and studied aspect of disease control in both the poultry industry and hospital environments (Dancer, 2011, Lin *et al.*, 2011, Bragg *et al.*, 2014). In hospital environments, bacterial clinical isolates are found to be less susceptible to common disinfectants compared to the culture collection type strains (Dancer 2011). This is usually due to confusion over disinfecting responsibilities in hospitals which lead to poor disinfection and the emergence of multi-drug resistant strains of nosocomial pathogens (Dancer 2011). In 2011, the fourth leading cause of disease in developed countries was nosocomial infections (Guggenbichler *et al.*, 2011). In the United States about 90000 fatalities occur every year due to hospital acquired infections. This number is assumed to be substantially worse in developing countries (Lin *et al.*, 2011). These resistant bacteria are not isolated in the hospital environment as they have also been identified in outpatients (Buffet-Bataillon *et al.*, 2012). The agricultural industry is under the same stress as it is perceived as normal to lose livestock as a result of multi-drug resistant bacterial infections. This is in spite of extensive use of antibiotics and disinfectants (Bragg *et al.*, 2014).

Antibiotic and disinfectant resistance is possibly the most threatening problem in human and animal healthcare. It remains unchecked, every surgery, whether a major or minor procedure, would have the potential to be a high risk and life threatening process (Davies *et al.*, 2011). In 2001, this prompted the experts at the WHO to release a document that explained preventative measures to slow down the emergence of highly resistant strains called the WHO global strategy for containment of antimicrobial resistance (Buffet-Bataillon *et al.*, 2012). It is important to note that any use of disinfectants whether they are used correctly or not, results in selective pressure for resistance (Tansirichaiya *et al.*, 2018).

4. IS THERE ANY LINK BETWEEN ANTIBIOTIC AND DISINFECTANT RESISTANCE?

There has been speculation of a link between resistances to different anti-microbial, known as cross-resistance. Templeton and co-workers (2009) found that antibiotic resistant in *Escherichia coli* strains were more chlorine tolerant than strains that were not resistant to antibiotics. Khan and co-workers (2016) went on to test the MIC values of chlorine tolerant bacteria against a variety of antibiotics. It was concluded that a significant (but weak) correlation between tolerance to the disinfectant and antibiotics did exist.

This cross-resistance may be due to a common underlying mechanism such as the multidrug membrane efflux pump or due to an accumulation of genetic mobile elements that confer resistance or tolerance to antibiotics and antimicrobials including disinfectants (Russel, 2002, Zhou *et al.*, 2017). Efflux pumps are the most prevalent mechanism of resistance to disinfectants and can be specific or used for many antimicrobial compounds in multidrug resistance pumps (Piddock, 2006, Hassan *et al.*, 2010). However, the proposed link is inconsistent and contradicting studies have also shown that this correlation is only present on rare occasions (Li *et al.*, 2012, Zhou *et al.*, 2017). Cross-resistance is highly variable, dependant on the type of antimicrobial and the bacterium. Simply because a bacterium is resistant to an antibiotic does not necessarily mean it will exhibit higher tolerance to disinfectants (Russel, 2002, Zhou *et al.*, 2017).

5. MECHANISMS OF RESISTANCE TO DISINFECTANTS

It is believed to be very rare for bacteria to develop resistance to disinfectants, as many disinfectants are complexes to have multiple target sites (Gnanadhas *et al.*, 2013). It is assumed that some microorganisms are prone to be more resistant than others spore forming bacteria (*Bacillus* spp., *Clostridium* spp.), naked viruses and gram-negative bacteria are usually more tolerant of disinfectants (Russel and Maillard 1997, Sattar 2007). This is due to differences in cellular structure and composition as well as the physiology of the bacterial cells (Bragg *et al.,* 2014), although variation within the nature of the disinfectants as well as the actual organisms do exist (Russell 1997). When resistance is discovered, understanding the mechanism is important so that we know how to overcome it. Intrinsic resistance refers to the natural ability of a bacterial cell to be less susceptible to an antimicrobial agent whereas acquired resistance is due to genetic changes by mutation or acquisition of genetic material through horizontal gene transfer (plasmids) (Russel and Maillard 1997). In order for any disinfectant or antimicrobial agent to be effective, it needs to reach its target site in high enough concentration. Therefore, many mechanisms work on decreasing the concentration of antimicrobials in the cell. The mechanisms of resistance or tolerance to disinfectants include:

5.1 Phenotypic Changes

In nutrient poor environments, bacteria form capsules, spores and/or biofilms which increase tolerance and resistance to numerous disinfectants (Brown and Gilbert, 1993, Gilbert *et al.,* 2002). The resistance of *Pseudomonas aeruginosa* against benzalkonium chloride increased 100-fold while it was a part of a biofilm (Gilbert *et al.,* 1990, Gilbert and McBain 2003). Numerous cases are recorded where bacteria have grown inside bottles of disinfectants and antiseptics themselves (Marrie and Costerton, 1981). *Serratia marcescens* has been found on multiple occasions in such cases (Marrie and Costerton, 1981). Biofilm formation was responsible for the prolonged survival of *S. marcescens* in 2% chlorhexidine (Marrie and Costerton, 1981). Cells within biofilms have different characteristics to free planktonic cells, this often includes an increased resistance to disinfectants. Various conditions contribute to resistance of biofilms, these include slow growth and/or induction of the *rpoS*-mediated stress response (Gilbert *et al.,* 2002). In addition, changes to biofilm architecture and physical or chemical variations of exo-polysaccharides can confer resistance by excluding disinfectants from the bacterial population. Bacteria in biofilms may exhibit a unique resistance profile dependant on the heterogeneous nature of the biofilm, with several mechanisms at work simultaneously (Mah and O'Toole 2001).

The envelope organisation and architecture differences in Gram-negative and Gram-positive bacteria, contribute to disinfectant susceptibility and tolerance. It has been suggested that Gram-negative bacteria are less susceptible to some disinfectants because of the presence of Outer Membrane Proteins (OMPs). OMPs impair the diffusion and uptake of some disinfectants into the cell (Murtough *et al.,* 2001). This is seen in *P. aeruginosa*, *S. marcescens* and *Proteus* sp. as an increase in tolerance is attributed to hindered uptake of disinfectants (Russel and Maillard 1997). A change in the charge of the bacterial cell surface can also act to hinder disinfectant uptake. Tansirichaiya and co-workers (2018) found that in the presence of *gal* E (involved in lipopolysaccharide production) the host cell surface of *E. coli* was more positively charged. This in turn impaired the binding of positively charged disinfectants. This is the first time a housekeeping gene such as *gal* E has been shown to confer resistance to QACs and other cationic antimicrobials (Tansirichaiya *et al.,* 2018).

5.2 Efflux Pumps

Efflux pumps are the most widespread mechanisms of disinfectant resistance. Efflux proteins actively pump the antimicrobial agents out of the cell, therefore lowering the intracellular concentration (McDonnell and Russell 1999, Piddock 2006, Buffet-Bataillon *et al.,* 2012). Broad-specificity efflux pumps are over expressed and are responsible for cross-resistance in many bacteria by exporting antimicrobials out of the cell (Nikaido 1998a). One example is the chromosomally encoded acrAB efflux pump in *E. coli* (and other gram-negative bacteria). This allows for intrinsic resistance to a variety of antibiotics (ciprofloxacin, tetracycline, fluoroquinolone, β-lactams) and other chemicals (deoxycholate, SDS, phenylethylalcohol, acriflavine, ethidium bromide) (Ma *et al.,* 1993, 1994; Nikaido 1998b).

5.3 Inactivation of Disinfectants

Along with cell wall modification and efflux pumps, another intrinsic method of resistance involves inactivation of disinfectants. Chlorhexidine, used for skin disinfection prior to surgery, is degraded by *Serratia marcescens* and *Pseudomonas aeruginosa*, posing a threat to bio-security in operating theatres (Ogase *et al.,* 1992). Strains of *Pseudomonas fluorescens* were found to have the ability to degrade didecyldimethyl ammonium chloride to decyldimethylamine and dimethylamine as intermediates. These strains were able to degrade QACs by a process of N-dealkylation, resulting in increased resistance to these disinfectants (Nishihara *et al.,* 2000).

The crystal structure of the functional unit or monomer of a representative protein of each group is shown where available. The MFS and SMR families crystal structures include the substrate represented as a green sphere. The orientation of these proteins within the cytoplasmic membrane (CM) and outer membrane (OM) of Gram-negative bacteria are shown. The ABC super family requires hydrolyses of ATP to provide energy for active transport, which is catalysed by the nucleotide binding domains (NBDs) of the transporter. The crystal structure of *S. aureus* Sav 1866 (pdb: 2HYD) functional unit of the ABC transporter consists of two trans-membrane domains (TMDs) and two NBDs. The five singlet system secondary active transporter families/super families are MFS, MATE, PACE, SMR and AbgT. Representative transport proteins are shown as crystal structures of *E. coli* MFS Mdfa (pdb:4ZOW), *V. cholerae* MATE NorM (NorM_VC) (pdb: 3MKT), *E. coli* SMR EmrE (pdb:3B5D) and *N. gonorrhoeae* AbgTM trF (pdb: 4R1l). The RND superfamily has multiple components spanning across the cytoplasmic membrane and outer membrane each highlighted in a different colour. Crystal structures of these components are of *E. coli* AcrB (pdb: 2DRD) representing the integral CM protein, AcrA (pdb: 2F1M) the membrane fusion protein (MFP), and TolC (pdb: 2VDE) the OM factor (OMF) of the RND efflux system are illustrated. (Chitzas and Brown 2017).

5.4 Target Alteration

Instead of inactivating the disinfectants themselves, target sites of the disinfectants can be altered to inhibit their activity. Triclosan, a component of surface disinfectants with antibacterial and antifungal properties inhibits fatty acid synthesis in bacteria. Triclosan targets Fab-I, an enoyl-acyl carrier protein reductase essential for bacterial fatty acid synthesis. A missense mutation in the Fab-I gene, alters the target site of the protein, and is responsible for resistance of the bacteria to triclosan (Heath *et al.,* 1999). Additional studies have shown that over expression of Fab-I in *E. coli* can confer resistance as well (Tansirichaiya *et al.,* 2018). In *Mycobacterium tuberculosis,* inhA (an analogue of fabI) is the target for ethionamide and isoniazid. However, a point mutation within this gene (inhA) confers resistance to this bacterium (Banerjee *et al.,* 1994). It has also been suggested that in *E. coli,* phenol resistance was seen to be conferred by changes in the outer membrane proteins as well as permeability changes in the lipid membrane (Zhang *et al.,* 2011). Altering the targets of typical phenols and therefore, their antimicrobial ability. This ability to modify the outer membrane is one reason why gram-negative bacteria have a greater advantage in increasing resistance and tolerance to antimicrobials and disinfectants (Murtough *et al.,* 2001). In addition, complex cell walls such as those in *Mycobacteria* have unique target sites for disinfectants and thus added targ*et al*teration opportunities. *Mycobacteria* possess water-filled porins that are necessary for the transport and activity of various components of disinfectants and sanitising products (methylenbisoxazolidine, isothiazolinones and octenidinedihydrochloride) (Frenzel *et al.,* 2011). The absence of these porins (MspA and MspaA-like porins) is responsible for reduced susceptibility to these compounds without compromising physiological function in the bacterium (Frenzel *et al.,* 2011).

5.5 Horizontal Gene Transfer

Horizontal gene transfer is a well-known mechanism for the development of disinfectant resistance. Resistance genes can be passed from one bacterium to another, plasmids can be transferred conferring resistance and transposable elements have been known to play a role as well (Cooksey 1987, Russel 1997). *Staphylococcus* species such as *S. aureus* decrease susceptibility to disinfectant compounds by plasmid-mediated mechanisms (Townsend *et al.,* 1983a, b, 1984). In addition to this, reduced susceptibility observed in *S. aureus* can be due to the acquisition of thesh-fab I allele from *Staphylococcus haemolyticus* by horizontal gene transfer (Ciusa *et al.,* 2012). The increased prevalence of these transferable elements in a bacterial population is due the positive selective pressure brought about by the presence of antimicrobial compounds present in disinfectants at sub MIC concentrations (Ciusa *et al.,* 2012). Bacteria in aquaculture environments have been shown to accumulate mobile genetic elements (integrating conjugative elements) which can confer resistance to QACs, heavy metals and rifampicin (Rodriguez-Blanco *et al.,* 2012). Additional genetic material coding for degradative enzymes was also reported.

These enzymes include o-phospho transferases, n-acetyl transferases and o-adenyl transferases capable of degrading antimicrobial compounds (Wright *et al.,* 1998).

5.6 QAC Resistance Genes

Quaternary Ammonium Compounds (QACs) are widely used antimicrobials with an effect against a broad range of microorganisms (Hegstad *et al.,* 2010). The bactericidal effect of QACs is brought about by disrupting the cell wall which subsequently leads to leakage of cytoplasm out of the cells. The systemic use of QACs in several products and across different industries makes the emergence of QAC resistance alarming. Various mechanisms of QAC resistance have been discussed already in this review. However, the genetic side of resistance in particular is becoming more important due to the sequencing technology available today.

Acquired resistance to QACs is well documented in *S. aureus* and *E. coli* (Lyon and Skurray 1987). The QAC genes in *Staphylococcal* species are mostly plasmid borne and encode for efflux pumps, in particular the membrane proteins that export harmful molecules conferring resistance against various compounds including disinfectants (QACs after which the genes are named) and some antibiotics (Langsrud *et al.,* 2003, Hegstad *et al.,* 2010, Wassenaar *et al.,* 2015). The QAC resistance genes were named as they were discovered, Qac A is the protein encoded by the gene *qac* A in the pSK1 family of plasmids (Wassenaar *et al.,* 2015). Staphylococcal multidrug-resistant gene qacA has been associated with resistance to monovalent and divalent cations, ethidium bromide, intercalating dyes, benzalkonium chloride and chlorhexidine (Littlejohn *et al.,* 1992). However, Qac B (encoded by *qac* B) differs as it confers resistance to QACs and intercalating dyes but little or no resistance to diamidines or chlorhexidine (Lyon and Skurray 1987, Grinius *et al.,* 1992, Grinius and Goldberg 1994). Both *qac* A and *qac* B require a transcriptional regulator *qac* R, the transcriptional repressor QacR is needed for expression of both these genes (Peters *et al.,* 2009).

The QacC protein is encoded for by the *qacC* gene (also known as smr), this protein has been identified as a homolog to an efflux protein in *E. coli* named EmrE (Grinius and Goldberg 1994). The protein name QacD is no longer in use as it was identified to be the same as

QacC and now is referred to as QacC (Wassenaar *et al.,* 2015). The next addition to the Qac resistance gene family was QacE, it was first described in *E. coli* of which functionally active derivatives (QacEΔ) have been identified in other bacterial genera (Kazama *et al.,* 1998). Subsequently, QacF was described, it was detected in several gram-positive bacteria but no homologs have been detected in *Staphylococcus* yet (Ploy *et al.,* 1998). QacG and QacH were detected at similar times both conferring high level ethidium bromide resistance with QacG additionally conferring resistance to benzalkonium chloride (Heir *et al.,* 1999a). QacJ was subsequently described and there after QacZ was detected in *Enterococcus faecalis* (Bjorland *et al.,* 2003, Braga *et al.,* 2011). There is a definite division within these QAC resistance genes and the proteins they encode for- QacA and QacB are members of the major facilitator super family (MFS); whereas the rest (QacC, QacE, QacF, QacG, QacH, QacJ and QacZ) are members of the small multidrug resistance (SMR) protein family (Heir *et al.,* 1999b, Wassenaar *et al.,* 2015). These two families evolved separately and therefore are genetically different; the functional similarities are an illustration of parallel evolution. QacE, QacF and QacZ are the only proteins (and corresponding genes) not present in *Staphylococcal* species (Gillings *et al.,* 2009, Zmantar *et al.,* 2011, Wassenaar *et al.,* 2015).

In clinical isolates the *qac* gene is commonly found in class 1 integrons, in environmental isolates the class 1 integrons contain diverse *qac* gene sets (Gillings *et al.,* 2009). The integrons collect resistance cassettes and can readily pick up new resistance determinants on plasmids, proving to be an efficient system to travel between different species (Hegstad *et al.,* 2010, Bragg *et al.,* 2014). The *qac* genes are not the only ones that encode for QAC resistance. In *E. coli,* five additional chromosomally-encoded genes confer QAC resistance namely: sugE, emrE, ydgE/ydgF and mdfA (Bay and Turner 2009). The gene mdfA like qacA/B, belongs to the MFS super family whereas the rest belong to the SMR family (Edgar and Bibi 1997, Bay and Turner 2009). Among all of these QAC resistance genes sugE (p), qacE, qacEΔ1, qacF and qacG are located on mobile genetic elements along with different antibiotic resistance genes allowing for the development of cross-resistance (Zou *et al.,* 2014).

Identifying and characterising genes that confer disinfectant resistance has never been more important. The advances in sequencing technology has allowed for a deeper understanding of resistance mechanisms and shed light on the possible ways to combat resistance.

6. THE USE OF SEQUENCING TECHNOLOGY IN ANALYSIS AND CHARACTERISATION OF RESISTANCE TO DISINFECTANTS

Next generation sequencing is new technology that can help with an old problem, as full genome sequencing is proving to be a vital tool in the race against bacterial resistance (Koser *et al.*, 2014). Sequencing technology has allowed for identification and analysis of bacteria without the need to culture them (Abdallah *et al.*, 2017). This allows for rapid identification and characterisation of bacteria that have never been cultured and/or that no one has ever seen before (Buermans and Dunnen 2014, Austin 2017). Whole genome sequencing is allowing for advances in new treatments due to the identification and characterisation of resistance mechanisms. In 2005, 454 pyrosequencing allowed for the identification of the target site of the anti-TB agent Bedaquiline (Andries*et al.*, 2005). It identified the F0 subunit of ATP synthase as the target region. Thereafter, hundreds of TB-causing *Mycobacterium* strains were sequenced to confirm this target gene was conserved. This information subsequently allowed for the approval of the first novel class of anti-TB agents in 40 years (Andries *et al.*, 2005, Zumla *et al.*, 2013).

In clinical settings, sequencing is used to distinguish between re-infection from a primary infection, or infection with a new bacterium. Comparative analysis of the genome sequence can confirm the presence of a new bacterium or one from a previous infection. This allows clinical staff to know if their environment or equipment is still harbouring the pathogen or if the staff themselves is carriers of the infection allowing for confirmation of effective biosafety measures or if improvements are needed (Niemann *et al.*, 2009, Eyre *et al.*, 2014, Koser *et al.*, 2014).

Whole genome sequencing could be especially important for surveillance of reoccurring diseases and those with high mutation rates and variants. Sequencing allows for identification of various resistant strains and tracking their movement between industries and countries (Bryant *et al.*, 2013, Nurwidya *et al.*, 2018). New mutations conferring resistance and mutation rates can be monitored as the genomes are sequenced over time (WHO 2014, Grundmann 2014). Systemic studies at the genomic, epigenomic and transcriptomic levels of more than 25 000 cancer genomes revealed a range of oncogenic mutations (Roukos 2010). This data is currently being used in the development of new cancer therapies.

Bacterial isolates that are of interest, due to their resistant capabilities can be analysed using sequencing techniques. Parallel sequencing with related or type strains followed by a bioinformatics-based approach will allow the sequences to be aligned. Once a reference alignment has been formed, the sequences can be examined for variants, mutations and the presence of antibiotic resistance genes or pathogenicity islands by comparison with the reference genome and those in public databases. Any variation found can be evaluated in a functional context and could point to specific disinfectant resistance mechanisms (Mardis 2008).

Safi and co-workers (2013) used whole genome sequencing to study the evolution of ethambutol resistance in *Mycobacterium tuberculosis* complex (the causative agent in TB). In doing so, it was discovered that resistance in this bacterium developed in a step-wise manner and a novel resistance mechanism was described for the first time (Safi *et al.*, 2013). Once the mechanism of resistance is known, this may help in deciding which treatment options are best to combat this particular strain (Koser *et al.*, 2014). Thus, personalised medicine may be enhanced with sequencing methods which may become a primary diagnostic tool in the future (Harris *et al.*, 2010, Grad *et al.*, 2014).

The advancements in sequencing methods provide us with a greater insight into bacterial genomics and the resultant resistance it confers. The emergence and re-emergence of multi-drug resistant bacterial infections is a growing public health problem, accentuated by a decrease in antimicrobial drug discovery (Pop and Salzberg 2008). The genomic data obtained is a useful tool in developing novel diagnostic assays, as well as new therapeutic options at a personal and population level (Punina *et al.*, 2015). The amount of data that can be generated with sequencing technology is infinite and so new software and algorithms will need to be developed in parallel. Genome sequencing is currently used to supplement real-time diagnostic methods in clinical settings (Pop and Salzberg 2008). In order for its use to become more systemic, integrated and easy-to-use bioinformatics toolsets for simple data interpretation needs to be established, as well as standards for verification and validation of results (Pop and Salzberg 2008). Nevertheless, this technology has the potential to direct day to day infection control and may even give rise to the field of personalised genomic epidemiology (Koser *et al.*, 2014, Punina *et al.*, 2015).

7. CONCLUSIONS

Disinfectants are currently used on a large scale in many industries. A few decades ago resistance to disinfectants was unheard of. Now resistance to disinfectants is becoming more prevalent and we need to start acting soon if we want to save our disinfectants. There are three things we can do to prevent this, firstly we need to reduce the overuse of disinfectants and therefore lower the selective pressure on bacteria to develop resistance (Murtough *et al.*, 2001). Efforts should be made to ensure that only disinfectants of high standard with proven efficacy should be used in the agricultural and health-care sectors. There are restrictions on the use of antibiotics but none on the use of other antimicrobials. Therefore, disinfectant stewardship should be implemented as it is with antibiotics. Secondly, infection control measures need to be put in place to reduce human to human and animal to human transmission (Hawkey and Jones 2009). Finally, emphasis must be put on research into alternative methods of bio-control and safety (Allen *et al.*, 2014).

Sequencing technology has allowed for the identification and characterisation of numerous genes and the subsequent conferred resistance. These genes can be passed between different strains, species and genera as demonstrated by the QAC resistance genes (Ciusa *et al.*, 2012). The prevalence of resistance to disinfectants is not a rare occurrence as it used to be. If we cannot look after the disinfectants we are currently using, we will be facing the same problems that we have now with antibiotics. However, with proper bio-security measures and the responsible use of disinfectants this can be avoided.

ACKNOWLEDGEMENTS

I'd like to acknowledge Prof. Bragg and Dr. Wouter van der Westhuizen for their guidance and support. Funding was supplied by the South African National Research Foundation (NRF).

REFERENCES

[1] Abdallah, R.A. *et al.* (2017). The impact of culturomics on taxonomy in clinical microbiology. *Antonie Van Leeuwenhoek* 110: 1327–1337.

[2] Adedeji, W. (2016). The treasure called antibiotics. *Ann Ib Postgr Med* 14: 56–57.

[3] Allen, H.K., Trachsel, J., Looft, T. and Casey, T.A. (2014). Finding alternatives to antibiotics. *Ann N Y Acad Sci* 1323: 91–100.

[4] Andries, K. *et al.* (2005). A diarylquinoline drug active on the ATP synthase of *Mycobacterium tuberculosis*. *Science* 307: 223–227.

[5] Austin, B. (2017). The value of cultures to modern microbiology. *Antonie Van Leeuwenhoek* 110: 1247–1256.

[6] Banerjee, A. *et al.* (1994). *inh*A, a gene encoding a target for isoniazid and ethionamide in *Mycobacterium tuberculosis*. *Science* 263: 227–230.

[7] Bay, D. and Turner R. (2009). Diversity and evolution of the small multidrug resistance protein family. *BMC Evol Biol* 9: 140.

[8] Bjorland, J., Steinum, T., Sunde, M., Waage, S. and Heir, E. (2003). Novel plasmid borne gene *qacJ* mediates resistance to quaternary ammonium compounds in equine *Staphylococcus aureus*, *Staphylococcus simulans*, and *Staphylococcus intermedius*. *Antimicrob Agents Chemother* 47: 3046–3052.

[9] Braga, T., Marujo, P., Pomba, C. and Lopes, M. (2011). Involvement and dissemination of the enterococcal small multidrug resistance transporter QacZ in resistance to quaternary ammonium compounds. *J Antimicrob Chemother* 66: 283–286.

[10] Bragg, R., Jansen, A., Coetzee, M. and van der Westhuizen, W. (2014). Bacterial Resistance to Quaternary Ammonium Compounds (QAC) disinfectants. *Adv Exp Med Biol* 808: 1–13.

[11] Brown, M. and Gilbert, P. (1993). Sensitivity of biofilms to antimicrobial agents. *J Appl Bacteriol* 74: 87S–97S.

[12] Brussow, H. (2007). Phage therapy: the western perspective. In: Bacteriophage: Genetics and Microbiology (ed. McGrath, S.) 159–192 (Caister Academic Press).

[13] Bryant, J.M. *et al.* (2013). Whole-genome sequencing to establish relapse or re-infection with *Mycobacterium tuberculosis*: a retrospective observational study. *Lancet Respir Med* 1: 786–792.

[14] Buermans, H. and Dunnen, J. (2014). Next generation sequencing technology: Advances and applications. *Mol Basis Dis.*, 1842: 1932–1941.

[15] Buffet-Bataillon, S., Tattevin, P., Bonnaure-Mallet, M. and Jovlivet-Gougeon, A. (2012). Emergence of resistance to antibacterial agents: The role of quaternary ammonium compounds—a critical review. *Int J Antimicrob Agents*, 39: 381–389.

[16] Canton, R., Gonzalez-Alba, J. and Galan, J. (2012). CTX-M enzymes: origin and diffusion. *Front Microbiol* 3: 277–283.

[17] CDC (2016). Life Expectancy. *National Centre for Health Statistics*, USA https://www.cdc.gov/nchs/fastats/life-expectancy.htm. (accessed 18 April 2018)

[18] Chan, B., Abedon, S. and Loc-Carrillo, C. (2013). Phage cocktails and the future of phage therapy. *Futur Microbiol* 8: 769–783.

[19] Chitsaz, M., Brown, M. (2017). The role played by drug efflux pumps in bacterial multidrug resistance. *Essays in Biochemistry* 61:127-139.

[20] Ciusa, M. *et al.* (2012). A novel resistance mechanism to triclosan that suggests horizontal gene transfer and demonstrates a potential selective pressure for reduced biocide susceptibility in clinical strains of *Staphylococcus aureus*. *J Antimicrob Agents* 40: 210–220.

[21] Cooksey, D. (1987). Tomato, Characterisation of a copper resistance plasmid conserved in copper-resistant strains of *Pseudomonas syringae* pv. *Appl Env Microbiol* 53: 454–456.

[22] Cotter, P., Ross, R. and Hill, C. (2013). Bacteriocins—a viable alternative to antibiotics? *Nat Rev Microbiol* 11: 95–105.

[23] D'Andrea, M., Arena, F., Pallecchi, L. and Rossolini, G. (2013). CTX-M-type beta-lactamases: A successful story of antibiotic resistance. *Int J Med Microbiol* 13: 305–317.

[24] Dancer, S. (2011). Hospital cleaning in the 21st century. *Eur J Clin Microbiol Infect Dis* 30: 1473–1481.

[25] Davies, S. (2011). Infections and the rise of antibiotic resistance. In Annual Report of the Chief Medical Officer. Volume II: Infections and the rise of antibiotic resistance.

[26] Davis, M. *et al.* (2015). Recent Emergence of Escherichia coli with Cephalosporin Resistance Conferred by bla CTX-M on Washington State Dairy Farms, 2015. *Appl Environ Microbiol* 81: 4403–4410.

[27] Edgar, R. and Bibi, E. (1997). MdfA, an *Escherichia coli* multidrug resistance protein with an extraordinarily broad spectrum of drug recognition. *J Bacteriol* 179: 2274–2280.

[28] Eyre, D.W. *et al.* (2014). Whole-genome sequencing demonstrates that Fidaxomicin is superior to Vancomycin for preventing reinfection and relapse of infection with *Clostridium difficile*. *J Infect Dis* 209: 1446–1451.

[29] Fox, B.G. and Blommel, P.G. (2009). Autoinduction of protein expression. In: Current Protocols in Protein Science. Chapter 5, Unit 5.23.

[30] Frenzel, E., Schmidt, S., Niederweis, M. and Steinhauer, K. (2011). Importance of porins for biocide efficacy against *Mycobacterium smegmatis*. *Appl Env Microbiol* 77: 3068–3073.

[31] Gilbert, P., Allison, D. and McBain, A. (2002). Biofilms *in-vitro* and *in-vivo*: do singular mechanisms imply cross-resistance. *J Appl Microbiol* 92: 98S–110S.

[32] Gilbert, P., Collier, P. and Brown, M. (1990). Influence of growth rate on susceptibility to antimicrobial agents: biofilms, cell cycle, dormancy, and stringent response. *Antimicrob Agents Chemother* 34: 1865–1868.

[33] Gilbert, P. and McBain, A. (2003). Potential impact of increased use of biocides in consumer products on prevalence of antibiotic resistance. *Clin Microbiol Rev.,* 16: 189–208.

[34] Gillings, M., Xuejun, D., Hardwick, S., Holley, M. and Stokes, H. (2009). Gene cassettes encoding resistance to quaternary ammonium compounds: a role in the origin of clinical class 1 integrons? *ISME J* 3: 209–215.

[35] Gnanadhas, D., Mararhe, S. and Chakravortty, D. (2013). Biocides-resistance, cross-resistance mechanisms and assessment. *Expert Opin Investig Drug* 22: 191–206.

[36] Gould, I. and Bal, A. (2013). New antibiotic agents in the pipeline and how they can overcome microbial resistance. *Virulence* 4: 185–191.

[37] Grad, Y., Kirkcaldy, R., Trees, D. *et al.* (2014). Genomic epidemiology of *Neisseria gonorrhoeae* with reduced susceptibility to cefixime in the USA: a retrospective observational study. *Lancet Infect Dis* 14: 220–226.

[38] Grinius, L., Dreguniene, G., Goldberg, E., Liao, C. and Projan, S. (1992). A staphylococcal multidrug resistance gene product is a member of a new protein family. *Plasmid,* 27: 119–129.

[39] Grinius, L. and Goldberg, E. (1994). Bacterial multidrug resistance is due to a single membrane protein which functions as a drug pump. *J Biol Chem* 269: 29998–30004

[40] Gross, M. (2013). Antibiotics in crisis. *Curr Biol* 23: 1063–1065.

[41] Grundmann, H. (2014). Towards a global antibiotic resistance surveillance system: a primer for a roadmap. *Ups J Med Sci* 119: 87–95.

[42] Guggenbichler, J., Assadian, O., Boeswald, M. and Kamer, A. 2016. Incidence and clinical impact of nosocomal infections associated with implantable biomaterials—catheters, ventilator associated pneumonia, urinary tract infections. *GMS Krakenhhyg Interdiszip,* 6(1).

[43] Harris, S.R. *et al.* (2010). Evolution of MRSA during hospital transmission and intercontinental spread. *Science* 327: 469–474.

[44] Hassan, K., Baltzer, S., Paulsen, I. and Brown, M. (2010). Pumping out biocides-cause for concern. *Microbiol Aust* 178–181.

[45] Hawkey, P. and Jones, A. (2009). The changing epidemiology of resistance. *J Antimicrob Chemother* 72: 3–10.

[46] Heath, R. *et al.* (1999). Mechanism of triclosan inhibition of bacterial fatty acid synthesis. *J. Biol Chem.,* 274: 11110–1114.

[47] Hegstad, K. *et al.* (2010). Does the wide use of quaternary ammonium compounds enhance the selection and spread of antimicrobial resistance and thus threaten our health? *Microl Drug Resist* 16: 91–104.

[48] Heir, E., Sundheim, G. and Holck, A. (1999). The qacG gene on plasmid pST94 confers resistance to quaternary ammonium compounds in Staphylococci isolated from the food industry. *J. Appl Microbiol,* 86: 378–88.

[49] Heir, E., Sundheim, G. and Holck, A. (1999). Identification and characterisation of quaternary ammonium compounds resistant Staphylococci from the food industry. *Int J Food Microbiol* 48: 211–219.

[50] Jia, B., Zhou, T. and Huang, A. (2008). Role of TMS5: Staphylococcal multidrug-efflux protein, *Qac Chin Med J. (Engl),* 121: 409–413.

[51] Johnson, R.P. *et al.* (2008). Bacteriophages for prophylaxis and therapy in cattle, poultry and pigs. *Anim Heal Res Rev.,* 9: 201–215.

[52] Kazama, H., Hamashima, H., Sasatsu, M. and Arai, T. (1998). Distribution of the antiseptic resistance gene qacE delta 1 in gram-positive bacteria. *FEMS Microbiol Lett.,* 165: 295–299.

[53] Khan, S., Beattie, T. and Knapp, C. (2016). Relationship between antibiotic- and disinfectant-resistance profiles in bacteria harvested from tap water. *Chemosphere* 152: 132–141.

[54] Koser, C., Ellington, M. and Peacock, S. (2014). Whole-genome sequencing to combat antimicrobial resistance. *Trends Genet,* 30: 401–407.

[55] Langsrud, S., Sundheim, G. and Brogman-Stahsen, R. (2003). Intrinsic and acquired resistance to quaternary ammonium compounds in food-related *Pseudomonas* spp. *J Appl Microbiol,* 95: 874–882.

[56] Li, S., Li, W. and Yan, P. (2012). The review of MRSA in resistance to disinfectant. *Chinese J. Disinfect,* 29: 223–226.

[57] Lin, Y., Stout, J. and Yu, V. (2011). Prevention of hospital acquired legionellosis. *Curr Opin Infect Dis,* 24: 350–356.

[58] Littlejohn, T. *et al.* (1992). Substrate specificity and energetics of antiseptic and disinfectant resistance in *Staphylococcus aureus. FEMS Microbiol Lett,* 74: 259–265.

[59] Lyon, B. and Skurray, R. (1987). Antimicrobial resistance of *Staphylococcus aureus*: genetic basis. *Microbiol Rev.,* 51: 88–134.

[60] Ma, D., Cook, D. and Alberti, M. (1993). Molecular cloning and charkaterisation of acrA and acrE genes in *Escherichia coli. J. Bacteriol,* 175: 6299–6313.

[61] Ma, D., Cook, D., Hearst, J. and Nikaido, H. (1994). Efflux pumps and drug resistance in gram-negative bacteria. *Trends Microbiol* 2: 489–493.

[62] Mah, T. and O'Toole, G. (2001). Mechanisms of biofilm resistance to antimicrobial agents. *Trends Microbiol,* 9: 34–39.

[63] Mardis, E.R. (2008). The impact of next-generation sequencing technology on genetics. *Trends Genet,* 24: 133–141.

[64] Marrie, T. and Costerton, J. (1981). Prolonged survival of *Serratia marcescens* in chlorhexidine. *Appl Env Microbiol,* 42: 1093–1102.

[65] McDonnell, G. and Russell, A. (1999). Antiseptics and Disinfectants: Activity, Action and Resistance. *Clin Microbiol Rev.,* 12: 147–179.

[66] Murtough, S., Hiom, S., Palmer, M. and Russel, A. (2001). Biocide rotation in the healthcare setting: is there a case for policy implementation. J Hosp Infect 48: 1–6.

[67] Niemann, S. *et al.* 2009. Genomic diversity among drug sensitive and multidrug resistant isolates of *Mycobacterium tuberculosis* with identical DNA fingerprints. *PLoS One,* 4: 7407.

[68] Nikaido, H. (1998). Antibiotic resistance caused by gram-negative multidrug efflux pumps. *Clin Infect Dis* 27: S32-41.

[69] Nikaido, H. (1998). The role of outer membrane and efflux pumps in the resistance of gram-negative bacteria. Can we improve drug access? *Drug Resist Updat* 1: 93–98.

[70] Nishihara, T., Okamoto, T. and Nishiyama, N. (2000). Biodegradation of didecyldimethylammonium chloride by *Pseudomonas fluorescens* TN4 isolated from activated sludge. *J. Appl Microbiol* 88: 641–647.

[71] Nurwidya F., Handayani D., Burhan E. and Yunus F. 2018. Molecular diagnosis of tuberculosis. *Chonnam Med* J 54: 1-9.

[72] Ogase, H., Nigai, I., Kameda, K., Kume, S. and Ono, S. (1992). Identification and quantitative analysis of degradation products of chlorhexidine with chlorhexidine-resistant bacteria with three-dimensional high performance liquid chromatography. *J Appl Bacteriol,* 73: 71–78.

[73] Oliveira, D., Tomasz, A. and De Lencastre, H. (2002). Secrets of success of a human pathogen: molecular evolution of pandemic clones of meticillin-resistant *Staphylococcus aureus. Lancet Infect,* 2: 180–189.

[74] Peters, K., Sharbeen, G., Theis, T., Skurray, R. and Brown, M. (2009). Biochemical characterisation of the multidrug regulator QacR distinguishes residues that are crucial for multidrug binding and induction of qacA transcription. *Biochem* 48: 9794–9800.

[75] Piddock, L. (2006). Clinically relevant chromosomally encoded multidrug resistance efflux pumps in bacteria. *Clin Microbiol Rev*19: 382–402.

[76] Piet, J., Kieran, J., Dara, L. and Avelino, A. 2016. *Listeria monocytogenes* in food: Control by monitoring the food processing environment. *African J. Microbiol Res* 10: 1–14.

[77] Ploy, M., Courvalin, P. and Lambert, T. (1998). Characterisation of In40 of *Enterobacter aerogenes* BM2688, a class 1 integron with two new gene cassettes, cmlA2 and qacF. *Antimicrob agents Chemother,* 42: 2557–2563.

[78] Pop, M. and Salzberg, S. (2008). Bioinformatics challenges of new sequencing technology. *Trends Genet,* 24: 142–149.

[79] Punina, N.V., Makridakis, N.M., Remnev, M.A. and Topunov, A.F. (2015). Whole-genome sequencing targets drug-resistant bacterial infections. *Hum Genomics* 9: 19–26.

[80] Rodriguez-Blanco, A., Lemos, M. and Osorio, C. (2012). Integrating conjugative elements (ICEs) as vectors of antibiotic, mercury and quaternary ammonium compounds resistance in marine aquaculture environments. *Antimicrob Agents Chemother,* 56: 2619–2626.

[81] Roukos, D. (2010). Complete genome sequencing and network modelling to overcome trastuzumab resistance. *Pharmacogenomics,* 11: 1039–1043.

[82] Russel, A. (2002). Introduction of biocides into clinical practsie and the impact on antibiotic-resistant bacteria. *Appl Microbiol,* 92: 121–135.

[83] Russel, A. (1997). Plamids and bacterial resistance to biocides. *J. Appl Microbiol* 83: 155–165.

[84] Russel, A. and Maillard, J. (1997). Microbial susceptibility and resistance to biocides. *ASM News,* 63: 481–487.

[85] Rutala, W. (1995). APIC Guidelines for selection and use of disinfectants. *Am J Infect Control,* 24: 313–342.

[86] Safi, H. *et al.* (2013). Evolution of high-level ethambutol-resistant tuberculosis through interacting mutations in decaprenylphosphoryl-β-D-arabinose biosynthetic and utilization pathway genes. *Nat Genet* 45: 1190–1197.

[87] Sattar, S. (2007). Hierarchy of susceptibility of viruses to environmental surface disinfectants: A predictor of activity against new and emerging viral pathogens. *J. AOAC* 90: 1655–1660.

[88] Sengupta, S., Chattopadhyay, M. and Grossart, H. (2013). The multi-faceted roles of antibiotics and antibiotic resistance in nature. *Front Microbiol,* 4: 13–26.

[89] Sidhu, M.H.S. and Holck, A. (2002). Resistance to quaternary ammonium compounds in food-related bacteria. *Microb Drug Resist,* 8: 393–399.

[90] Spellberg, B. and Gilbert, D. (2014). The future of antibiotics and resistance: A ribute to a career of leadership by John Barlett. *Clin Infect Dis,* 59: S71–S75.

[91] Tansirichaiya, S. *et al.* (2018). Reduced susceptibility to antiseptics is conferred by heterologous housekeeping genes. *Microb Drug Resist,* 24: 105–112.

[92] Templeton, M., Oddy, F., Leung, W. and Rogers, M. (2009). Chlorine and UV disinfection of ampicillin-resistant and trimetoprim-resistant *Escherichia coli. Can J. Civ. Eng.,* 36: 889–894.

[93] Townsend, D., Ashdown, N., Greed, L. and Grubb, W. (1984). Transposition of gentamicin resistance to staphylococcal plasmids encoding resistance to cationic agents. *J. Antimicrob Chemother,* 14: 115–124.

[94] Townsend, D., Grubb, W. and Ashdown, N. (1983). Genetics of drug resistance in methicillin-resistant *Staphylococcus aureus* from Australian hospials. *J. Hosp Infect,* 4: 331–337.

[95] Townsend, D., Greed, L., Ashdown, N. and Grubb, W. (1983). Plasmid-mediated resistance to quatarnery ammonium compounds in methicillin-resistant *Staphylococcus. Med J Aust.,* 2: 310.

[96] Udikovic-Kolic, N., Wichmann, F., Broderick, N. and Handelsman, J. (2014). Bloom of resident antibiotic-resistant bacteria in soil following manure fertilization. *Proc Natl Acad Sci. USA,* 111: 15202–15207.

[97] van Panhuis, W.G. *et al.* (2013). Contagious Diseases in the United States from 1888 to the Present. *N Engl. J. Med* 369: 2152–2158.

[98] Ventola, C. (2015). The antibiotic resistance crisis. *Pharm Ther,* 40: 277–283.

[99] Wassenaar, T., Ussery, D., Nielsen, L. and Ingmer, H. (2015). Review and phylogenetic analysis of qac genes that reduce susceptibility to quaternary ammonium compounds in *Staphylococcus* species. *Eur J. Microbiol Immunol*, 5: 44–61.

[100] WHO (2018). World Health Organization-Listeriosis in South Africa.

[101] http://www.who.int/csr/don/28-march-2018-listeriosis-south-africa/en/ (Accessed: 24th July 2018).

[102] WHO (2014). World Health Organization 2014. The evolving threat of antimicrobial resistance: Options for action. In: WHO Publications. doi: ISBN 9789241503181.

[103] Wright, G., Berghuis, A. and Mobashery, S. (1998). Aminoglycoside antibiotics structures, functions and resistance. *Adv. Exp. Med. Biol.*, 456: 27–69.

[104] Zaffiri, L., Gardner, J. and Toledo-Pereyra, L. (2012). History of antibiotics. from salvarsan to cephalosporins. *J. Investig Surg* 25: 67–77.

[105] Zhang, D., Li, H., Lin, X., Wang, S. and Peng, X. (2011). Characterization of outer membrane proteins of *Escherichia coli* in response to phenol stress. *Curr Microbiol*, 62: 777–783.

[106] Zhou, Q. *et al.* (2017). Is the potential Interconnection between Antibiotic susceptibility and disinfectant resistance a threat to food safety. *Agric Sci Technol*18: 1033–1037.

[107] Zmantar, T., Kouidhi, B., Miladi, H. and Bakhrouf, A. (2011). Detection of macrolide and disinfectant resistance genes in clinical *Staphylococcus aureus* and coagulase-negative staphylococci. *BMC Res Notes* 4: 453.

[108] Zou, L. *et al.* (2014). Presence of disinfectant resistance genes in *Escherichia coli* isolated from retail meats in the USA. *J. Antimicrob Chemother,* 69: 2644–2649.

[109] Zumla, A., Nahid, P. and Cole, S.T. (2013). Advances in the development of new tuberculosis drugs and treatment regimens. *Nat Rev. Drug Discov,* 12: 388–404.

National Antimicrobial Resistance Surveillance in Nepal: What the Laboratory-based Data Reveals?

Nisha Rijal*, **Runa Jha and Jyoti Acharya**

National Public Health Laboratory, Department of Health Services, Ministry of Health and Population, Teku, Kathmandu, Government of Nepal

E-mail: *nisharijal1@gmail.com

ABSTRACT: Antimicrobial resistance (AMR) is one of the major challenges faced in public health in this century and a threat to current and future medical advances. AMR problem is more frequent in a developing country like Nepal where indiscriminate, inadequate and inappropriate use of antimicrobials and self-medication are quite common. In Nepal, lab based AMR surveillance program started in 1999 with nine laboratories in its initial phase monitoring six pathogens of interest. Currently, 21 hospitals/laboratories are included in AMR surveillance monitoring ten organisms of interest. The data received from all participating laboratories on ten organisms of interest (i.e. Salmonella spp., Shigella spp., V. cholerae, S. pneumoniae, H. influenzae, N. gonorrhoea, ESBL E. coli, MRSA, MDR Klebsiella and Acinetobacter) during January - December, 2017 were compiled and analyzed for specimen type, age and sex wise distribution and resistance pattern for each organism. Of the total 2337 isolates of interest reported in 2017, the most common organism was ESBL E. coli (834, 35.6%) followed by Salmonella (464, 19.8%), MDR Klebsiella (349, 14.9%), Methicillin Resistant Staphylococcus aureus (327,13.9%), MDR Acinetobacter spp. (213,9.1%) and others. The age and sex-wise distribution revealed diversity in predominant age group and sex depending on the organisms. The antimicrobial susceptibility patterns also varied widely among various pathogens, however, increase in multidrug resistance and fluctuation in antimicrobial susceptibility was a common phenomenon. Continuous monitoring of changing anti-biogram is not only essential for empiric purposes but also provides a picture of current trend in antimicrobial resistance and help the governing authorities to take necessary measures to combat the emerging resistance.

Keywords: MRSA, AMR-Nepal, Resistance Pattern, Anti-biogram, AMR-surveillance.

1. INTRODUCTION

The antimicrobial resistance phenomenon has become a global concern as geographic borders among countries and continents have become less distinct. The development of AMR is a natural phenomenon in microorganisms, and is accelerated by the selective pressure exerted by use and misuse of antimicrobial agents in humans and animals (WHO 2014). Factors such as over the counter sales of antibiotics, under consumption of a specified dose of antimicrobial, lack of well-equipped laboratory and diagnostic facilities inclusion of antibiotics as growth promoters in plants and animals and lack of awareness on ill effects of AMR in health and economy has compounded the problem in developing countries (Ayukekbong *et al.*, 2017). Since AMR has emerged in a wide range of infectious agents in all regions of the world, controlling a single agent by a single agency would be impossible. A multi sectoral collaboration between various agencies and countries is inevitable (WHO 2016). In Sept. 2011, the Government of Nepal has also committed to ***'Jaipur Declaration'*** to combat anti-microbial resistance where they have agreed to advocate for a multi-disciplinary approach by all sectors of the government, with the private health sector providing desired information and following national guidelines (Tacconeli *et al.,* 2018). Surveillance is the key to measure antimicrobial resistance and is expected to provide relevant information regarding changes in microbial populations, early detection of resistant strains, prompt notification and investigation of outbreaks. Surveillance findings are also essential to monitor the appropriateness of therapy guidelines, antibiotic formulary, antibiotic stewardship programmes, public health interventions, infection control policies, and antimicrobial development (Shakya *et al.*, 2014).

In Nepal, laboratory based AMR surveillance started since 1999 with technical support from ***International Centre for Diarrhoeal Disease Research (ICDDR, B)*** Bangladesh. Back then, it had 9 laboratories as sentinel sites and monitored six pathogens and since 2006, the programme is supported by the World Health Organization. At present, the surveillance has 21 sentinel sites and monitors ten pathogens (*Salmonella* spp., *Shigella* spp., *V. cholerae, S. pneumoniae, H. influenzae, N. gonorrhoea*, ESBL *E.coli*, Methicillin Resistant *S. aureus*, MDR *Klebsiella* and *Acinetobacter*). This paper attempts to compile and analyze the lab based data generated on antimicrobial susceptibility from various sites and disseminate a comprehensive national finding on the status of bacterial resistance.

2. MATERIALS AND METHODS

2.1 Study Sites

The study was carried out in 21 sentinel sites, out of which only 15 laboratories (National Public Health Laboratory (NPHL), Patan Hospital, Kanti Children's Hospital, B.P. Koirala Institute of Health Sciences, Western Regional Hospital, Manipal Teaching Hospital, United Mission Hospital, Lumbini Zonal Hospital, Dhulikhel Hospital, Central Veterinary Laboratory, Kathmandu Model Hospital, KIST Medical College, Koshi Zonal Hospital, Bayal Pata Hospital, Paropkar Maternity and Women's Hospital and Sukraraj Tropical and Infectious Disease Hospital) reported isolates of interest.

2.2 Study Duration

The study was carried out between January–December, 2017.

2.3 Study Population

All inpatients/outpatients attending respective hospitals were referred for culture of sample.

2.4 Laboratory Methods

Adequate samples were collected in leak proof container or suitable transport medium and transported to the laboratory at the earliest possible. The samples were processed following Standard Microbiological Techniques. All the positive samples (*i.e.* samples yielding significant bacterial growth) were further processed for identification by standard techniques and antimicrobial susceptibility by Kirby Bauer disc diffusion method using antibiotic discs recommended by CLSI specific for the target organism.

2.5 Data Sharing, Analysis and Compilation

A monthly report from all participating laboratories was sent to National Public Health Laboratory in a standard format which included a unique patient ID, demographic details such as age/sex, location of patient, ward, sample type, organism isolated, and date of isolation and anti-microbial susceptibility pattern of only selected organisms. The data thus obtained from each laboratory was compiled on a monthly basis and a monthly report was dispatched to each participating laboratory. At the end of the year, the compiled data was analyzed on the basis of rate of isolation, demographic status, sample type and a comprehensive annual AMR report was prepared, the findings of which were disseminated to clinicians, laboratory workers and pharmacists through dissemination workshops, EIC materials and through web.

3. RESULTS AND DISCUSSION

A total of 2337 isolates of surveillance interest were reported in the year 2017. Majority of the isolates were reported from Patan hospital (Figure 1).

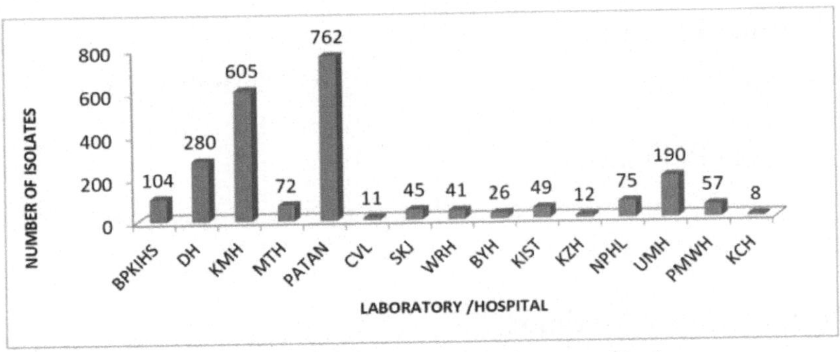

Figure 1: Hospital/Lab Wise Distribution of Isolates

Of the total isolates reported in 2017, the most common organism was by ESBL *E. coli* (834, 35.6%) followed by *Salmonella* (464, 19.8%), MDR *Klebsiella* (349, 14.9%), Methicillin Resistant *Staphylococcus aureus* (327, 13.9%), and MDR *Acinetobacter* spp (213, 9.1%). Similarly, 74 isolates of *Streptococcus pneumoniae* (3.1%), 62 *Shigella* isolates (2.6%). 6 isolates of *Haemophilus influenzae* (0.25%) and only 8 isolate of *Neisseria gonorrhea* were reported from AMR participating laboratories (Figure 1).

The isolation and reporting of fastidious pathogen is low and is mainly due to the low number of samples from which they can be isolated and also the difficulty to isolate and preserve it from other overgrowing commensals and pathogens.

Sex-wise distribution of the isolates demonstrated that the proportion of isolates from male was higher for *Salmonella, H. influenzae* and *N. gonorrhoea* where as isolates of ESBL *E. coli,* Methicillin Resistant *Staphylococcus aureus* and *Shigella* spp. were reported higher from females. Equal rate of infection among males and females was seen in MDR *Klebsiella, Acinetobacter* spp. and *Streptococcus pneumoniae* (see Figure 2). Various studies show variation in age and sex depending upon the sample and organisms isolated.

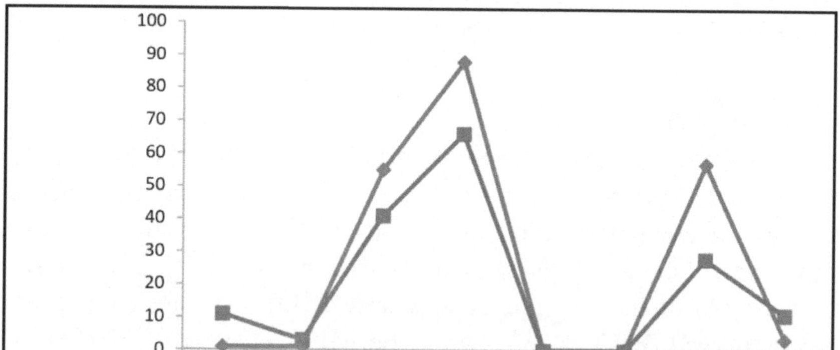

	Amp	Ts	Cip	Na	C	CRO	LEV	Azt
S.typhi	1	1	55	88	0.2	0.3	57	3.6
S.paratyphi A	11	3	41	66	0	0	28	11

S.typhi S.paratyphi A

Figure 2: Comparative Antibiogram of Typhoidal *Salmonella*

3.1 *Salmonella* spp.

Among the *Salmonella* isolates reported, 371 (79.9%) were *Salmonella enterica* subsp. *Typhi,* 67 isolates (14.4%) were *Salmonella enterica subsp. Paratyphi* A, 7 isolates (1.5%) were *Salmonella enterica* subsp. *enterica* serovar Typhimurium and 19 (4.09%) isolate were other *Salmonella* species. Similar enteric fever based studies made in Nepal shows predominance of *Salmonella Typhi* (Shakya *et al.,* 2014) while some shows *S. Paratyphi* to be predominant (Acharya *et al.,* 2012, Pokharel *et al.,* 2009). Our results showed 78% of the isolates were reported from age group of 20–29 years followed by age group of 10–19 years. Similar results were obtained by Mathura *et al.* (2005) where the mean age of patients was 26.17 years. However, some studies conducted by Pokharel *et al.* (2006) and Sharma *et al.* (2006) in Nepal showed that patients of 16–30 years age group were most common (Acharya *et al.*, 2012, Pokharel *et al.*, 2009, Gupta *et al.*, 2013). 48% of *Salmonella* were fluoroquinolone resistant in our study which is much higher than reported by Shakya *et al.* (2010) which shows only 14% resistance (Shakya *et al.*, 2016). Increasing resistance against fluoroquinolone has been reported from various parts of the world and has become a major concern. Resistance to Ciprofloxacin was higher in *Salmonella Typhi* as compared to *S. Paratyphi* in our study and so was true in the study by Gupta *et al.* (2013) in India. The comparative anti-biogram of typhoidal *Salmonella* is shown in Figure 2.

3.2 *Shigella* spp.

Among 62 *Shigella* spp, 59 were reported from stool sample, whereas 3 were reported from liver of poultry. *S. flexneri* was predominant. Majority of cases were reported from children less than 10 years in both sexes. The result is similar to that reported by Shakya *et al.* (2016), 37% of the *Shigella* isolates were MDR. Studies by Das *et al.* (2013) showed 30–32% *Shigella* to be MDR (Das *et al.*, 2013). The antibiogram of *Shigella* spp is represented in Figure 3. Eighty five percent of *Shigella* was resistant to Nalidixic acid and 54% to Ciprofloxacin which is higher than that reported by AMR program 2016 data summary report from Philippines which states a cumulative rate of resistance at 13.7% against ciprofloxacin (ARSP 2016). The higher resistance, may be because, in the past decade ciprofloxacin was used as mainstay therapy for treatment of shigellosis.

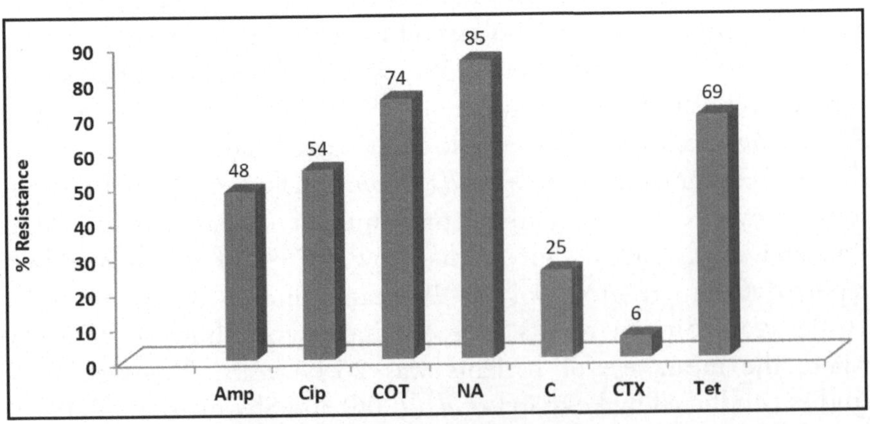

Figure 3: Antibiogram of *Shigella* spp.

3.3 *Streptococcus pneumoniae*

A total of 74 *Streptococcus pneumoniae* isolates were reported of which 71% were recovered from sputum. Infection was predominant in females and most of the cases were reported from patients 60 years above age group. Since most of the studies focused in Nepal are based on meningitides in children, relevant papers demonstrating *S. pneumoniae* in adult population are limited. Our results are similar to a study conducted in Western Nepal by Easow *et al.* (2011) which states that patients above 19 years (96.2%) were most common and majority (94.2%) of isolates were recovered from sputum. Our results showed 41% resistance against Cotrimoxazole which was much lower to that reported in 2008. Shakya and Adhikari (2012) stated that 8% isolates were resistant to ceftriaxone in 2007 which declined to 4% by 2008 (Shakya and Adhikari, 2012). In our study we found no isolate resistant to ceftriaxone. This may be due to gradual decrease in resistance. Only 9% isolates were resistant to penicillin which is much lower than that (26.7%) reported by Easow *et al.* (2011) but higher than that reported by Sherchand *et al.* who reported only 1% isolates as intermediate to penicillin (Easow *et al.*, 2011, Sherchan *et al.* (2015). The detailed antibiogram is represented in Figure 4.

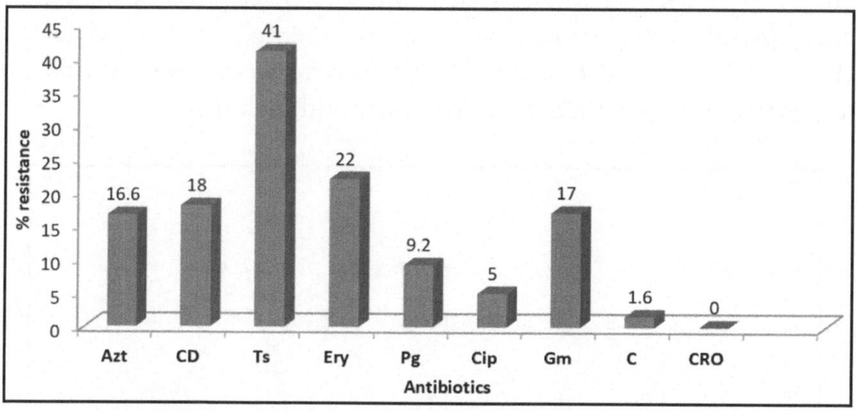

Figure 4: Antibiogram of Streptococcus pneumoniae

3.4 *Haemophilus influenzae* and *Neiserria gonorrhoae*

Only 6 isolates of Haemophilus *influenzae* and 8 isolates of *Neisseria gonorrhea* were reported. All the *Haemophilus* isolates were reported from sputum. 66% of the reported isolates were resistant towards

Cotrimoxazole and 50% to Chloramphenicol which is much higher to that reported by AMR surveillance data from Philippines (ARSP 2016). All the eight *Neisseria gonorrhoeae* were recovered from urethral swab, pus and penile discharge of males. Only one isolate was resistant to Ciprofloxacin which is much lower to that reported from other gonococcal surveillance studies (Kirkcaldy *et al.* 2016, Shakya *et al.*, 2017). This high resistance rates among fastidious pathogens observed by us may be because of low number of isolates.

3.5 Extended Spectrum Beta-Lactamases (ESBL) Producing *E. coli*

A total of 834 Extended Spectrum Beta-Lactamases producing *E. coli* isolates were reported of which 89% was recovered from urine with male to female ratio being 1:1.8. Infection rate was higher in the age group of 70-79 years in case of males and in age group of 20-29 years among females. Similar studies made in Nepal also showed high incidence in females (Shakya *et al.*, 2017). The antibiotic susceptibility of *E. coli* to various antibiotics is shown in Figure 5. In our study, 98% isolates were susceptible to Imipenem followed by Amikacin and Nitrofurantoin. Study made by Elsayed *et al.* (2017) also shows 96% sensitivity to Imipenem (Elsayed *et al.*, 2017). Reports of high resistance against carbapenems have been documented by Ye *et al.* (2015) whose study revealed that 83.3% enterobacteriaceae were resistant to imipenem, meropenem and ertapenem simultaneously.

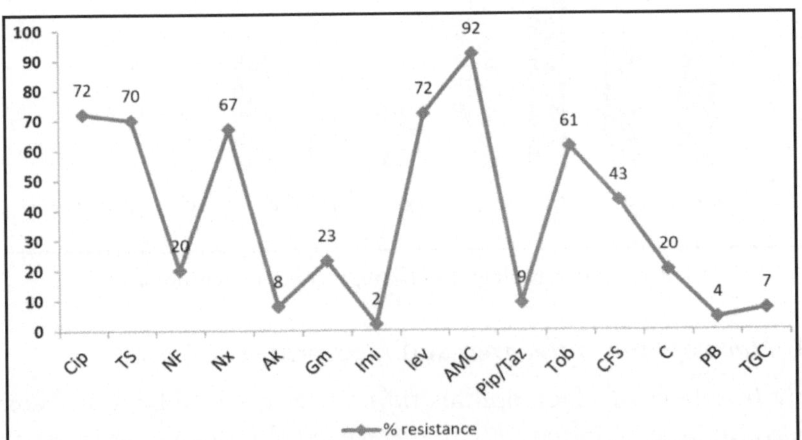

Figure 5: Antibiogram of Extended Spectrum Beta-Lactamases producing *E. coli*

3.6 Methicillin-Resistant Staphylococcus Aureus (MRSA)

A total of 327 Methicillin-Resistant Staphylococcus Aureus were reported of which 77% isolates were recovered from pus. Infection was higher in males of 1–9 years age group whereas in females it was reported from 20–29 years age group. However, study made by Ike *et al.* (2016) demonstrated high carriage rate in females (56.5%) than male (43.5%) and was highest in individuals of 20–30 years of age (57.65%). The difference may be due to geographical variation and also because the study was community based. The antimicrobial susceptibility profile shows 98% resistance against Penicillin G. Study made by Tiwari *et al.* (2009) in Western Nepal showed 100% resistance against Penicillin. Resistance against other antibiotics in his study was similar to our results. All the isolates were sensitive to tigecycline and Linezolid in our study. Figure 6 depicts the antibiotic susceptibitlity pattern of MRSA.

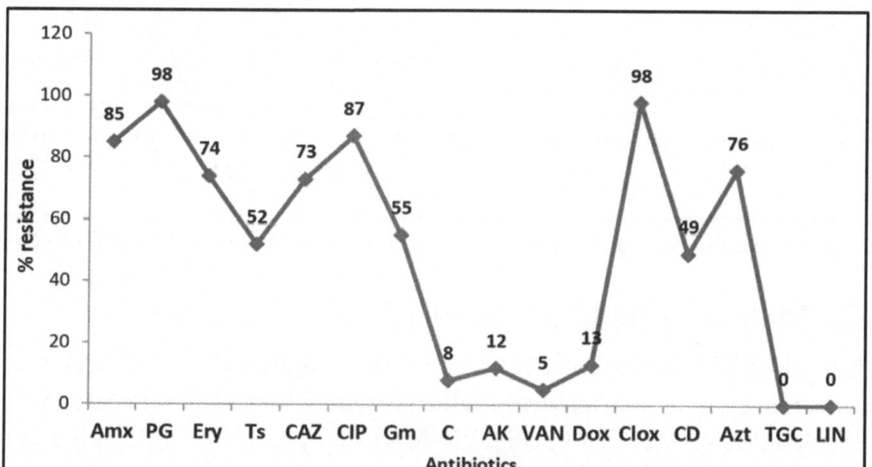

Figure 6: Antibiogram of Methicillin-Resistant Staphylococcus Aureus

3.7 Multidrug-Resistant (MDR) Klebsiella Isolates

A total of 349 Multidrug-Resistant Klebsiella isolates were reported of which 81% were *Klebsiella pneumoniae,* 3% were *K. oxytoca* and 16% were *Klebsiella* spp. Higher isolation rate of *K. pneumoniae* as compared to *K. oxytoca* has also been documented by Rath and Padhy (2014) and Biradar and Roopa (2015). Most of the isolates were recovered from Urine (93%) followed by sputum and pus. Of the total *Klebsiella* spp., majority were reported from age group of 20–29 years followed by 30–39 years in both sexes. The *Klebsiella* isolates in our study were 100% resistant to Ampicillin/Sulabcatam followed by 95% to Ceftazidime, 87% to Cotrimoxazole, 82% to Ciprofloxacin and others. All the isolates were sensitive to Polymyxin B. The resistance rates of various antibiotics against *Klebsiella* varied widely among various studies depending upon the site of origin and species (Babakhani *et al.*, 2014, Apondi *et al.*, 2016). The antibiogram of *Klebsiella* isolates against various antibiotics is demonstrated in Figure 7.

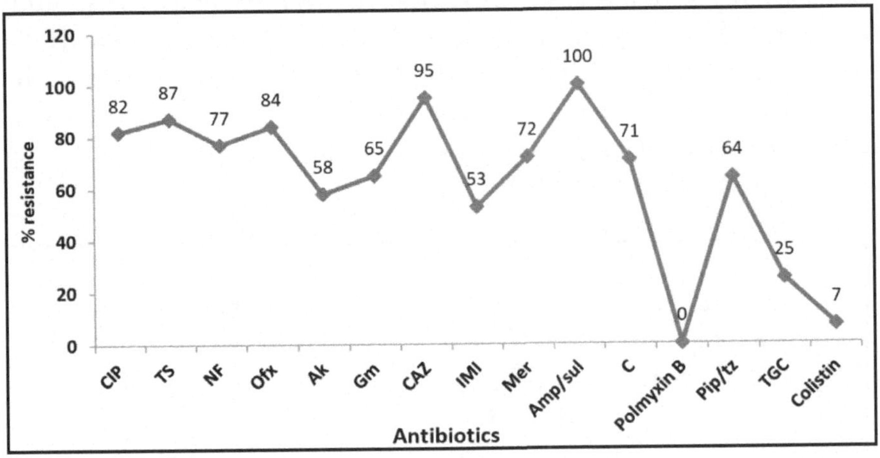

Figure 7: Antibiotic Susceptibility Pattern of Multidrug-Resistant Klebsiella Isolates

3.8 Multidrug-Resistant (MDR) Acinetobacter spp.

A total of 213 isolates of Multidrug-Resistant Acinetobacter spp. were reported from 9 participating laboratories. Most of the isolates were recovered from sputum (32.8%) followed by pus (26.7%). Study made by Sohail *et al.* (2016) also showed high isolation of *Acinetobacter* from respiratory samples. Infection was higher in patients aged 20–29 years in females and 60–69 years in male. More than 90% isolates were resistant to Ceftazidime, Meropenem, Ampicillin/Sulabactam and Amoxycillin. Above 80% are resistant to Ofloxacin, Chloramphenicol, Cotrimoxazole and Ciprofloxacin. Around 70% isolates were susceptible to Tigecycline. Colistin was most susceptible (97%) towards MDR *Acinetobacter* spp. Similar result of antibiogram was obtained by Batarseh *et al.* (2016) which showed high resistance rates of *A. baumannii* for ceftriaxone, cefotaxime and ticarcillin (100%), ceftazidime, cefepime and piperacillin (98.3%), imipenem (97.4%), piperacillin/tazobactam (96.6%), quinolones (94.8%), ampicillin/sulbactam (89.7%), gentamicin, (87.9%), tobramycin and tetracycline (76.7%) and trimethoprim/sulfamethoxazole (75.9%), but lower for minocycline (26.7%) and colistin (1.7%). The antibiogram of Multi-drug-Resistant Acinetobacter spp. is depicted in Figure 8.

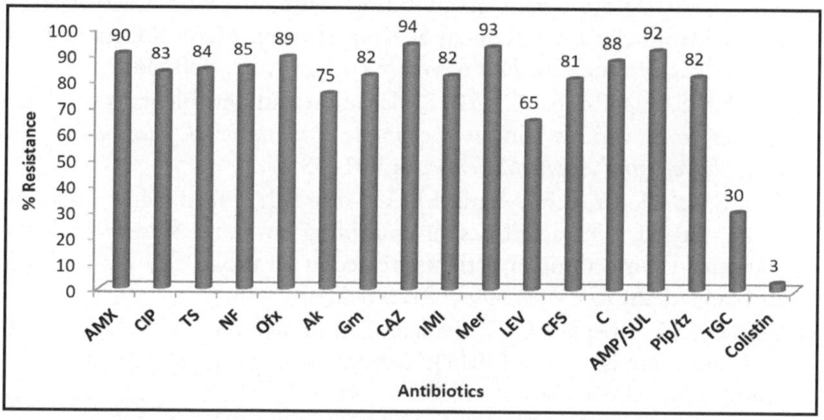

Figure 8: Antibiogram of Multidrug-Resistant Acinetobacter spp.

4. CONCLUSION

Continuous monitoring of pathogens for their changing pattern in antimicrobial resistance not only helps to guide the empiric therapy but also plays a major role in predicting the future and help in better management of antimicrobial resistance.

ACKNOWLEDGEMENTS

We acknowledge the entire laboratory staff of the AMR sentinel sites for their active contribution in isolation, identification and sharing of the data.

REFERENCES

[1] Acharya, A., Nepal, H.P., Gautam, R. and Shrestha, S. (2012). Enteric Fever pathogens and their antimicrobial susceptibility in Chitwan, Nepal. *Journal of Chitwan Medical College* 1: 26–30.

[2] Apondi, O.E., Oduor, O.C., Gye, B.K. and Kipkoech, M.K. (2016). High prevalence of multi drug resistant *Klebsiella pneumonia* in a tertiary hospital in western Kenya. *Afr J Infect Dis,* 10: 89–95.

[3] ARSP (2016). Antimicrobial Resistance Surveillance Program, Data Summary Report-2016, Antimicrobial Resistance Surveillance Reference Laboratory (ARSRL), Research Institute for Tropical Medicine, Department of Health Philippines, p. 19.

[4] Babakhani, S., Derikvand, S.S., Nazer, M.N. and Kazemi, M.J. (2014). *Bulletin of Environment, Pharmacology and Life Sciences,* 3: 149–154.

[5] Batarseh, A., Al-Sarhan, A., Maayteh, M., Al-Khatirei, S. and Alarmouti, M. (2016). Anti-biogram of multidrug resistant *Acinetobacter baumannii* isolated from clinical specimens at King Hussein Medical Centre, Jordan: a retrospective analysis. *East Mediterr Health J.*, 21: 828–34.

[6] Biradar, S. and Roopa, C. (2015). Isolation and anti-biogram of *Klebsiella* species from various clinical specimens. *International Journal of Current Microbiology and Applied Sciences,* 4: 991–995.

[7] Das, S.K., Klontz, E.H., Azmi, I.J., Ud-din, A.I., Chisti, M.J., Afrad, M.K. *et al.* (2013). Characteristics of multidrug resistant *Shigella* and *Vibrio cholerae* O1 infections in patients treated at an urban and a rural hospital in Bangladesh. *ISRN Microbiol* 21: 3915.

[8] Easow, J.M., Joseph, N.M., Shankar, P.R., Rajamony, A.P., Dhungel, B.A. and Shivananda, P.G. (2011). *Streptococcus pneumoniae* infections in western Nepal. *Southeast Asian J. Trop Med Public Health,* 42: 912–919.

[9] Elsayed, T.I., Ismail, H.A., Elgamal, S.A., *et al.* (2017). The occurrence of multidrug resistant *E. Coli* which produce ESBL and cause urinary tract infections. *J Appl Microbiol Biochem,* 1: 2–8.

[10] Gupta, V., Singla, N., Bansal, N., Kaistha, N., Chander, J. (2013). Trends in the antibiotic resistance patterns of enteric fever isolates—A three year report from a tertiary care centre. *Malays J. Med Sci* 20: 71–75.

[11] Ike, B., Ugwu, M.C., Ikegbunam, M.N., Nwobodo, D., Ejikeugwu, C., Gugu, T. and Esimone, C.O. (2016). Prevalence, anti-biogram and molecular characterization of community-acquired Methicillin-Resistant *Staphylococcus aureus* in AWKA, Anambra Nigeria. *Open Microbiol J.,* 10: 211–221.

[12] James, A., Ayukekbong, M.N. and Andrew, N.A. (2017). The threat of antimicrobial resistance in developing countries: causes and control strategies. *Antimicrobial Resistance and Infection Control* 6: 47.

[13] Kirkcaldy, R.D., Harvey, A., Papp, J.R., Del Rio, C., Soge, O.O., Holmes, K.K., Hook, E.W., Kubin, G., Riedel, S., Zenilman, J., Pettus, K., Sanders, T., Sharpe, S. and Torrone, E. (2016). *Neisseria gonorrhoeae* antimicrobial susceptibility surveillance—The Gonococcal isolate surveillance project, 27 Sites, United States, 2014. *Morbidity and Mortality Weekly Report Summary,* 65:1–19.

[14] Mathura, K.C., Chaudhary, D., Simkhada, R., Pradhan, M., Shrestha, P. and Gurubacharya, D.L. (2005). Study of clinical profile and antibiotic sensitivity pattern in culture positive typhoid fever cases. *Kathmandu University Med J.,* 3: 376–379.

[15] Pokharel, B.M., Koirala, J., Dahal, R., Mishra, S.K., Khadga, P.K. and Tuladhar, N.R. (2006). Multidrug-resistant and extended-spectrum beta-lactamase (ESBL)—producing *Salmonella enterica* (serotypes Typhi and Paratyphi A) from blood isolates in Nepal: surveillance of resistance and a search for newer alternatives. *Int J Infect Dis.* 10: 434–438.

[16] Pokharel, P., Rai, S.K., Karki, G., Katuwal, A., Vitrakoti, R. and Shrestha, S.K. (2009). Study of enteric fever and anti-biogram of *Salmonella* isolates at a Teaching Hospital in Kathmandu Valley. *Nepal Med College Journal* 11: 176–178.

[17] Rath, S. and Padhy, R.N. (2014). Prevalence of two multidrug-resistant *Klebsiella* species in an Indian teaching hospital and adjoining community. *Journal of Infection and Public Health,* 7: 496–507.

[18] Shakya, G. and Adhikari, B. (2012). Ten-years surveillance of antimicrobial resistance pattern of *Streptococcus pneumoniae* in Nepal. *African Journal of Microbiology Research,* 6: 4233–4238.

[19] Shakya, G., Acharya, J., Adhikari, S. and Rijal, N. (2016). Shigellosis in Nepal: 13 years review of nationwide surveillance. *J. Health Popul Nutr,* 35: 36.

[20] Shakya, G., Shrestha, S., Adhikari, S., Upadhaya, B.P., Rijal, N. and Sharma, S. (2014). Changing trend of antimicrobial susceptibility pattern of *Salmonella* spp. in different regions of Nepal: A six years surveillance study. *Academia Journal of Microbiology Research* 2: 71–78.

[21] Shakya, P., Shrestha, D., Maharjan, E., Sharma, V.K. and Paudyal, P. (2017). ESBL production among *E. coli* and *Klebsiella* spp. causing urinary tract infection: A Hospital Based Study. *Open Microbiol J.,* 11: 23–30.

[22] Sharma, N.P., Peacock, S.J., Phumratanaprapin, W., Day, N., White, N. and Pukrittayakamee, S. (2006). A hospital-based study of bloodstream infections in febrile patients in Dhulikhel Hospital, Kathmandu University Teaching Hospital, Nepal. Southeast. *Asian J. Trop Med Public Health,* 37: 560–568.

[23] Sherchan, J.B., Tandukar, S., Shrestha, L., Rai, G.K., Sharma, A., Gami, F.C., Rijal, B. and Ohara, H. (2015). *Streptococcus pneumoniae* among children with clinical meningitis in Nepal. *Journal of Institute of Medicine,* 37: 47–52.

[24] Sohail, M., Rashid, A., Aslam, B., Waseem, M., Shahid, M. and Akram, M. (2016). Antimicrobial susceptibility of *Acinetobacter* clinical isolates and emerging anti-biogram trends for nosocomial infection management. *Rev Soc Bras Med Trop,* 49: 300–304.

[25] Tacconeli, E., Sifakis, F., Harbath, S., Schrijver, R., Mourik, M., Voss, A. *et al.* (2018). Surveillance for control of antimicrobial resistance. *THE LANCET Infectious disease* 18: e99–106.

[26] Tiwari, H.K., Das, A.K., Sapkota, D., Sivrajan, K. and Pahwa, V.K. (2009). Methicillin resistant *Staphylococcus aureus*: prevalence and anti-

biogram in a tertiary care hospital in western Nepal. *J. Infect Dev Countries,* 3: 681–684.

[27] Who (2016). Workshop on National Action Plan on Antimicrobial Resistance for Developing Countries: Focusing on resistance emanating from antibiotic use in food animals. 10–11 November 2016, New Delhi, India, http://www.searo.who.int/regional_director/speeches/2016/combat ing_antimicrobial/en

[28] Ye, Y., Xu, L., Han, Y., Chen, Z., Liu, C. and Ming, L. (2018). Mechanism for carbapenem resistance of clinical enterobacteriaceae isolates. *Experimental and Therapeutic Med.,* 15: 1143–1149.

ABBREVIATIONS

Abbreviation	*Antibiotic Name*
Amp	Ampicillin
Cip	Ciprofloxacin
C	Chloramphenicol
TS/COT	Cotrimoxazole
NA	Nalidixic Acid
CRO	Ceftriaxone
Ofx	Ofloxacin
Ery	Erythromycin
Tet	Tetracycline
AZT	Azithromycin
NF	Nitrofurantoin
PG	Penicillin G
AMC	Amox-Clav
GM	Gentamicin
CTX	Cefotaxime
NX	Norfloxacin
CAZ	Ceftazidime
Lev	Levofloxacin
Pip/Tz	Piperacillin/Tazobactam
Cef/sul	Cefoperazone/Sulbactam
Mer	Meropenem

Abbreviation	*Antibiotic Name*
Imi	Imipenem
Ak	Amikacin
Lin	Linezolid
Dox	Doxycycline
TGC	Tigecycline
Tob	Tobramycin
PB	Polymyxin B
CL	Colistin
AMR	Antimicrobial Resistance
AST	Antibiotic Susceptibility Testing
ESBL	Extended Spectrum ß-Lactamases
ICDDR/B	International Centre for Diarrhoeal Disease Research/Bangladesh
MDR	Multi-Drug Resistant
MRSA	Methicillin Resistant Staphylococcus aureus
NPHL	National Public Health Laboratory
USAID	United States Aid for International Development
WHO	World Health Organization

Multidrug Resistant *Escherichia coli* Bearing *qnrS* and *blaTEM* Genes from Healthy Chickens in Zaria, Kaduna State, Nigeria

Jibril Yakubu Jibril[1*], Kabir Junaidu[2], J.K.P. Kwaga[2], Sanusi Atiku Abubakar[3], Bakari Nana Aisha[2], Sameer Sanusi Yusuf[4] and Mamman Mohammed[5]

[1]Department of Agricultural Biotechnology, National Biotechnology Development Agency, Abuja, Nigeria

[2]Department of Veterinary Public Health and Preventive Medicine, Faculty of Veterinary Medicine, Ahmadu Bello University, Zaria, Nigeria

[3]Kano State Ministry of Agriculture and Natural Resources, Kano, Nigeria

[4]Centre for Biotechnology Research and Training, Ahmadu Bello University, Zaria, Nigeria

[5]Department of Veterinary Pharmacology and Toxicoloogy, Faculty of Veterinary Medicine, Ahmadu Bello University, Zaria, Nigeria
&
Africa Centre of Excellence for Neglected Tropical Diseases and Forensic Biotechnology, Ahmadu Bello University, Zaria, Nigeria

E-mail: *jyjsul@gmail.com

ABSTRACT: Fluoroquinolones and Extended Spectrum Beta-Lactam (ESBL) antibiotics represent important classes of antibiotics in both veterinary and clinical practice. Given their efficacy in treating infections resistant to other drugs, these drugs have enjoyed widespread and often indiscriminate use. This study was aimed at determining the occurrence of antibacterial resistant E. coli from poultry and their potential for disseminating plasmid-borne qnr and ESBL gene variants known to increase MICs by multiple folds and facilitate selection of higher level resistance. Twenty-five (25%) of farms registered with the poultry clinic of the Veterinary Teaching Hospital, Ahmadu Bello University, Zaria within the year 2014 were selected using simple random sampling technique. 30% E. coli isolation rate was recorded. The highest resistance to fluoroquinolone was seen with enrofloxacin (49.3%) followed by ofloxacin and ciprofloxacin with 33.3 and 22.7% respectively, while 58.7% were resistant to nalidixic acid. About 93.3, 80 and 76% were respectively resistant to amoxicillin/clavulanic acid, streptomycin and sulphamethoxazole/trimethoprim. Resistance to chloramphenicol, tetracycline and gentamicin also respectively accounted for 68, 65.3 and 52%. Least resistance was however seen with doxycycline (10.7%), nitrofurantoin (6.7%), ceftriaxone (6.7%), imipenem (2.7%) and cefepime (1.3%). Multidrug resistance was seen in 92% of the isolates tested, all of which had a MAR index of ≥ 0.20. qnrS and blaTEM were found in eighteen (24%) of the E. coli isolates. This study has demonstrated a high prevalence of multidrug resistance among E. coli isolates from poultry in Zaria. It further revealed the existence of fluoroquinolone and ESBL resistance genes. Consequently, it is recommended that poultry farmers use antibiotics as therapeutic only when necessary in accordance with good drug use and management practices under the prescription of a certified veterinarian.

Keywords: Fluoroquinolones, ESBL, E. coli, qnrS, blaTEM, PCR.

1. INTRODUCTION

Antimicrobial drugs revolutionized the control of bacterial infections. This success, however, is threatened by the emergence of resistance to commonly used antimicrobial drugs. Livestock production systems are implicated as reservoirs and vectors in the dissemination of resistance (Cohen and Tauxe, 1986; Van den Bogaard and Stobberingh, 2000; De Leener, 2005; Nsofor *et al.*, 2013).

Beta-lactam antibiotics and fluoroquinolones (FQs) represent important groups of antimicrobials in both clinical and veterinary practice and have been jointly categorized among the critically important antimicrobial agents by the Food and Agriculture Organization (FAO), World Organization for Animal Health and World Health Organization (WHO) (FAO, 2008). While the beta-lactam antibiotics constitute one of the oldest and most popular classes of antibiotics in current clinical practice (Suarez and Gudiol, 2009; Page, 2012), FQs are the only class of antimicrobial agents in clinical use that directly inhibit bacterial DNA synthesis (Dalhoff, 2012) and have become first- and second-line antimicrobials of choice for acute respiratory, enteric and urinary tract infections as well as serious systemic infections such as bacteremia (Thielman and Guerrant, 2004; Jamison, 2006).

Resistance may be expressed as spontaneous mutations in response to sublethal concentrations or via mobile genetic elements including plasmids (Dalhoff, 2012; Hopkins, Davies, and Threlfall, 2005; Strahilevitz *et al.*, 2009). Recent emergence of plasmid-mediated quinolone resistance mechanisms due to variant *qnr* genes and those coding for extended spectrum beta-lactamases is exacerbating the problem of resistance in clinically important bacteria.

Nigeria like other developing countries is faced with the problem of low level of biosecurity in poultry production, requiring massive use of antibiotics to prevent and control infections and suppress the effect of management stress factors. Consequently *there is profuse and uncontrolled use of antimicrobial agents in poultry production in Nigeria, including both FQs and cephalosporins* (Van den Boogard *et al.*, 2010; Paolo *et al.*, 2008) raising serious concern for public health. Several studies have documented the rapid emergence of resistant zoonotic bacteria following use of antimicrobial agents in poultry (Van den Bogaard and Stobberingh, 2000; Randall *et al.*, 2003; Johnson *et al.*, 2007). These organisms are shed in feaces (Aarestrup, 2000) and can contaminate poultry tissues at slaughter or the environment during processing and through poultry wastes. A key organism of concern is *Escherichia coli* which is known to share resistant determinants with other enteric and environmental bacteria (Paulo *et al.*, 2013). Contamination with *E. coli* is used as indicator for carcass and environmental contamination with other zoonotic enteric bacteria including *Salmonella* species. Given its population and a dynamic economy, Nigeria's poultry production is not only a means of livelihood but an important contributor to the country's GDP. It is therefore necessary for the public health system in Nigeria to monitor the emergence and spread of AMR in human-agriculture interface. This study was therefore aimed at determining the occurrence of antibacterial resistant *E. coli* from poultry and their potential for disseminating plasmid-borne *qnr* and ESBL gene variants.

2. MATERIALS AND METHODS

2.1 Study Area

Zaria is a major city in Kaduna State, North Western Nigeria. The city is located within latitudes 11°04'N 7°42'E and longitudes 11.067°N 7.700°E with an altitude of 675 meters above sea level and a total land area of 300 km^2 (100 sq mi). It also has a density of 1,400/km^2 and a population of 975,228 according to the 2006 Census.

2.2 Sampling

A total of 250 cloacal swabs from healthy chickens involving 20 (25%) poultry farms randomly selected from Ahmadu Bello University Zaria Veterinary Teaching Hospital Poultry clinic register for the year 2014 were collected between October 2014 and January 2015. Samples were transported to the laboratory in Stuart's medium.

2.3 Isolation of Escherichia coli from Chickens

Samples were enriched in lauryl sulfate tryptose (LST) broth containing MUG (4-methylumbelliferyl-B-D-glucuronide) for 24 h at 37°C. Eosin Methylene Blue (EMB) agar plates were used to grow samples over night at 37°C and then characterized using conventional biochemical tests including Tripple Iron Sugar (TSI), Simmons citrate, Urea, Sulfate-Indole-Motility (SIM) and Methyl Red (MR), Voges Proskauer (VP). Isolates were then confirmed using Microbact GNB 12E (Oxoid, Uk) according to manufacturer's instructions.

2.4 Antimicrobial Susceptibility Testing

Susceptibility of *E. coli* isolates to 15 different antibiotics was determined using the Kirby-Bauer method and interpreted according to Clinical and Laboratory Standards Institute (CLSI, 2011) criteria. These antibiotics include Nitrofurantoin 300 μg, Nalidixic acid 30 μg, Ciprofloxacin 5 μg, Ofloxacin 5 μg, Enrofloxacin 5 μg, Cefepime 30 μg, Ceftriaxone 30 μg, Doxycycline 30 μg, Imipenem 10 μg, Tetracycline 30 μg, Streptomycin 25 μg, Sulphamethoxasole/Trimethoprim 25 μg, Chloramphenicol 30 μg, Gentamicin 10 μg, and Amoxicillin/Clavulanic acid 30 μg.

2.5 Detection of qnr and ESBL Genes

E. coli isolates resistant to fluoroquinolones from the disk diffusion test were subjected to a multiplex polymerase chain reaction to detect three (3) *qnr* (*qnr* A, *qnr* B, *qnr* S) and three (3) ESBL genes (*bla*-SHV, *bla* TEM, and *bla* CTX-M). Following purification on EMB plates, bacterial DNA was isolated from a 24-hour culture in LB broth using ZR Fungal/Bacterial DNA MiniPrep™ (from Zymo Research, USA) according to manufacturer's instructions. The isolated DNA was then ran on agarose gel electrophoresis to confirm presence of genomic

DNA. Extracted DNA was subjected to PCR in a 10 µl reaction mixture containing 3.5 µl PCR Master Mix, 2.5 µl water, 3.0 µl DNA and 0.5 µl each of forward and reverse primers used. Amplification was conducted as follows: Initial denaturation for 2 min at 94°C and 15 cycles of amplification consisting of 15 seconds at 93°C, 30 seconds at 65°C and 30 seconds at 72°C; another 40 cycles of amplification consisting of 15 seconds at 95°C, 30 seconds at 55°C and 30 seconds at 72°C before a final extension at 72°C for 5 minutes.

3. RESULTS

Microbact® confirmed 75 (30%) of the two hundred and fifty (250) cloacal swabs collected and tested as positive for E. coli. Out of these, only 1 (1.3%) was susceptible to all antibiotics tested, while 3 isolates (4%) were sensitive to only 1 antibiotic, 2 (2.7%) others were sensitive to only 2 antibiotics of different classes. The other 69 (92%) isolates were sensitive to three or more antibiotics. Highest percentage resistance to fluoroquinolone was seen with enrofloxacin (49.3%) followed by ciprofloxacin and ofloxacin with 33.3 and 22.7% resistance respectively while 58.7% of the isolates were resistant to nalidixic acid. Up to 93.3, 80 and 76% of the isolates were resistant to amoxicillin/clavulanic acid, streptomycin and sulphamethoxazole/ trimethoprim respectively. Resistance to chloramphenicol, tetracycline and gentamicin was exhibited by 68%, 65.3% and 52% of the isolates respectively. Least resistance was seen with doxycycline (10.7%), nitrofurantoin (6.7%), ceftriaxone (6.7%), imipenem (2.7%) and cefepime (1.3%) (Table 1). The isolates also showed 53 different patterns of resistance with AMC, S, TE, SXT, C and AMC, NA, CIP, OFX, ENR, CN, S, TE, SXT, C antibiograms combinations having the highest frequencies with 6 isolates each (Table 2) followed by AMC, NA, CN, S, TE, SXT, C exhibited by 4 isolates. Multidrug resistance, as defined by the joint committee of the European Centre for Disease Prevention and Control and the Centers for Disease Control and Prevention (Magiorakos *et al.*, 2012), was observed in 69 (92%) isolates of the 75 isolates. Out of these, 5 (6.7%) isolates had MAR index of < 0.20 (cut off value) while only 1 (1.3%) isolate had a value > 0.80. Whereas 21 (28%) isolates had values between 0.20 and 0.39, 33 (44%) and 14 (18.7%) isolates had values between 0.40 and 0.59 and 0.60 and 0.79 respectively (Table 3).

Eighteen (18) fluoroquinolone resistant isolates were tested for three variants each of qnr (qnrA, qnrB and qnrS) and ESBL (blaTEM, blaSHV and blaCTX-M) genes in a multiplex PCR using primers reported by Cattoir *et al.*, 2007 and Monstein *et al.*, 2007 respectively. All 18 isolates tested showed amplification of genes corresponding to *qnr S* (428bp) (Plate I) and *blaTEM* (445bp) (Plate II). This indicates co-carriage of these genes in all the 18 fluoroquinolone resistant *E. coli* isolates tested. These isolates also showed multidrug resistance (Table 4). The other genes *qnr*A, *qnr*B and *bla*SHV and *bla*CTX-M were not detected in the multiplex PCR.

Table 1: Percentage Resistance of *E. coli* Isolates to Various Antimicrobials

Sl. No.	*Antibiotic*	*Number Resistant*	*Percentage (%)*
1.	Nalidixic Acid	44	58.7
2.	Ciprofloxcin	25	33.3
3.	Ofloxacin	17	22.7
4.	Enrofloxacin	37	49.3
5.	Amoxicillin/Clavulanic Acid	70	93.3
6.	Ceftriaxone	5	6.7
7.	Cefepime	1	1.3
8.	Imipenem	2	2.7
9.	Gentamicin	39	52.0
10.	Streptomycin	60	80.0
11.	Tetracycline	49	65.3
12.	Doxycycline	8	10.7
13.	Sulphamethoxasole/trimethoprim	57	76.0
14.	Chloramphenicol	51	68.0
15.	Nitrofurantoin	5	6.7

Table 2: Antibiotic Resistance Patterns of *E. coli* Isolates from Cloaca of Commercial Poultry in Zaria to 15 Antimicrobials

Sl. No.	*Pattern*	*Number of Isolates*	*Sl. No.*	*Pattern*	*Number of Isolates*
1.	NIL	1	28.	AMC, NA, CN, S, SXT, C	1
2.	AMC	2	29.	AMC, NA, CN, S, TE, C	1
3.	ENR	1	30.	AMC, NA, ENR, S, SXT, C	1
4.	AMC, SXT	1	31.	AMC, NA, ENR, S, TE, SXT	1
5.	AMC, TE	1	32.	AMC, NA, OFX, ENR, CN, TE	1
6.	AMC, CN, C	1	33.	AMC, CN, S, TE, DO, SXT, C	2
7.	AMC, CIP, ENR, TE	1	34.	AMC, NA, CIP, OFX, ENR, CN, C	1
8.	AMC, CIP, S, SXT	1	35.	AMC, NA, CN, S, TE, DO, SXT	1
9.	AMC, CN, S, TE	1	36.	AMC, NA, CN, S, TE, SXT, C	4
10.	AMC, NA, ENR, C	1	37.	AMC, NA, ENR, CN, S, TE, C	1
11.	AMC, NA, S, SXT	1	38.	AMC, CIP, OFX, ENR, CN, S, TE, C	1
12.	AMC, NA, S, TE	1	39.	AMC, NA, CIP, ENR, CN, S, SXT, C	2
13.	AMC, S, SXT, C	2	40.	AMC, NA, CIP, ENR, S, TE, SXT, C	1
14.	AMC, S, TE, SXT	1	41.	AMC, NA, CIP, OFX, ENR, CN, S, SXT	1
15.	AMC, CN, S, TE, SXT	1	42.	AMC, NA, CIP, OFX, ENR, CN, SXT, C	2
16.	AMC, ENR, S, TE, SXT	1	43.	AMC, NA, CN, S, TE, DO, SXT, C	1

(Table contd...)

(Table 2 contd...)

Sl. No.	*Pattern*	*Number of Isolates*	*Sl. No.*	*Pattern*	*Number of Isolates*
17.	AMC, NA, CIP, ENR, SXT	1	44.	AMC, NA, ENR, CN, S, TE, DO, SXT	1
18.	AMC, NA, ENR, S, SXT	1	45.	AMC, NA, ENR, S, TE, SXT, C, F	1
19.	AMC, S, TE, DO, SXT	1	46.	AMC, NA, CIP, ENR, CN, S, TE, SXT, C	2
20.	AMC, S, TE, SXT, C	6	47.	AMC, NA, CIP, OFX, ENR, CN, S, TE, C	1
21.	AMC, CN, S, TE, SXT, C	3	48.	AMC, NA, CIP, OFX, ENR, S, TE, SXT, C	1
22.	AMC, CRO, FEP, S, IPM, SXT	1	49.	NA, CIP, OFX, ENR, S, TE, SXT, C, F	1
23.	AMC, CRO, NA, S, SXT, F	1	50.	AMC, NA, CIP, OFX, ENR, CN, S, TE, SXT, C	6
24.	AMC, CRO, S, TE, SXT, C	1	51.	AMC, NA, OFX, ENR, CN, S, TE, DO, SXT, C	1
25.	AMC, ENR, S, TE, SXT, C	1	52.	AMC, NA, CIP, ENR, CN, S, TE, SXT, C	1
26.	AMC, NA, CIP, ENR, S, C	1	53.	AMC, CRO, NA, CIP, ENR, CN, S, TE, SXT, C, F	1
27.	AMC, NA, CIP, ENR, TE, C	1	54.	AMC, CRO, NA, ENR, CN, S, TE, DO, IPM, SXT, C, F	1
	Total	**36**			**39**

NA: Nalidixic Acid, CIP: Ciprofloxacin, OFX: Ofloxacin, ENR: Enrofloxacin, AMC: Amoxicillin/Clavulanic Acid, CRO: ceftriaxone, FEP: Cefepime, IPM: Imipenem, CN: Gentamicin, S: Streptomycin, TE: Tetracycline, DO: Doxycycline, SXT: Sulphamethoxasole/Trimethoprim, C: Chloramphenicol, F: Nitrofurantoin.

Table 3: Multiple Antibiotic Resistance Indices (MAR) of various *E.coli* Isolates

Sl. No.	*MAR Index Category*	*Number of Isolates*
1.	<0.20	5
2.	0.20–0.39	21
3.	0.40–0.59	33
4.	0.60–0.79	14
5.	>0.80	1
	Total	74

Table 4: Resistance Profile of *E. coli* Isolates Carrying *qnr*S and *bla*TEM

Isolate No.	*Resistance Profile*	*MAR Index*
1.	AMC, CIP, ENR, TE	0.27
2.	NA, CIP, OFX, ENR, S, TE, SXT, C, F	0.60
3.	AMC, NA, ENR, S, SXT	0.33
4.	AMC, CRO, NA, CIP, ENR, CN, S, TE, SXT, C, F	0.73
5.	AMC, NA, ENR, S, TE, SXT	0.40
6.	AMC, NA, CIP, ENR, S, C	0.40
7.	AMC, NA, CIP, OFX, ENR, CN, S, TE, SXT, C	0.67
8.	AMC, NA, ENR, S, SXT, C	0.40
9.	AMC, NA, CIP, OFX, ENR, CN, S, SXT	0.53
10.	AMC, NA, CIP, ENR, CN, S, TE, SXT, C	0.60
11.	AMC, NA, CIP, ENR, CN, S, TE, SXT, C	0.60
12.	AMC, NA, CIP, OFX, ENR, CN, S, TE, C	0.60
13.	AMC, CRO, NA, ENR, CN, S, TE, DO, IPM, SXT, C, F	0.80
14.	AMC, NA, ENR, CN, S, TE, C	0.47
15.	AMC, NA, OFX, ENR, CN, S, TE, DO, SXT, C	0.67
16.	AMC, NA, CIP, ENR, SXT	0.33
17.	AMC, NA, CIP, OFX, ENR, CN, SXT, C	0.53
18.	AMC, CIP, S, SXT	0.27

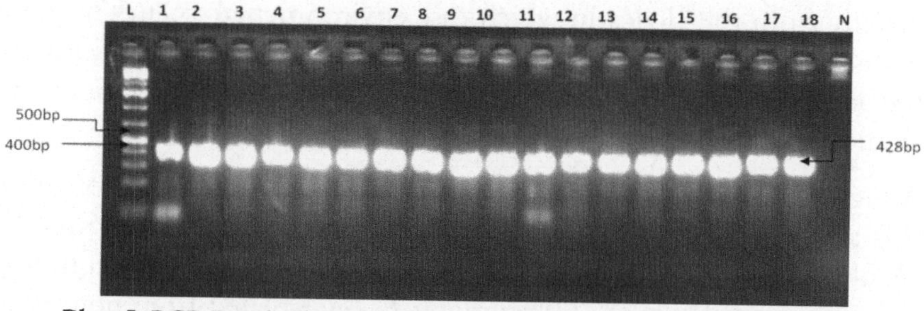

Plate I: PCR Result Showing *qnr* S Amplification in 18 *E. coli* Isolates. 1–18 Isolates Number, L= DNA Ladder, N = Negative Control

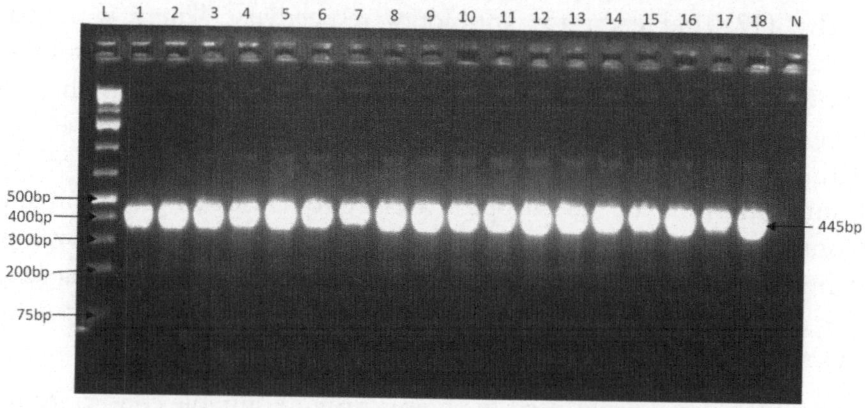

Plate II: PCR result showing *bla* TEM Amplification in 18 *E. coli* isolates. L = DNA ladder, N = Negative control

4. DISCUSSION

In this study, E. coli was isolated from cloaca of 75(30%) of 250 chickens sampled in Zaria for the presence of antibacterial resistant pathogens as confirmed by Microbact® 12E. Even though the bacterium is considered a part of the normal flora of the gastrointestinal tract of poultry, other factors such as survival in sample, presence of microbial inhibitors such as drugs consumed by the bird amongst other factors might have limited the isolation rate. These 75 isolates were confirmed using Microbact® 12E, a biochemical identification system developed mainly for the confirmation of human isolates which might have returned some E. coli from poultry as negative. This study's prevalence is, however, less than the 81% prevalence obtained by Adenipekun *et al.* (2015). The higher prevalence may be explained by the fact that Adenipekun et al. collected voided faecal samples not directly from the cloaca in which case environmental isolates may not be ruled out and also in that study, they used a different identification test kit (VITEK_2 System and the VITEK 2GN identification card from bioMe´rieux, Durham, NC).

A high prevalence (92%) of multidrug resistance was also recorded in this study. This finding is in close agreement with that reported by Ogunleye *et al.* (2008) and Melton (2014) who recorded a multidrug resistance prevalence of 100% among *E. coli* isolated from poultry in Abeokuta and Yola respectively. Fifteen (15) antibiotics were used in this study, while Ogunleye *et al.* (2008) and Melton (2014) used twelve (12) and eight (8) antibiotics respectively. There is a wide variation in types and panels of antibacterials used for susceptibility testing of *E. coli* for resistance surveillance and monitoring, which is bound to yield different outcomes. Similarly there is variation in the definition of multidrug resistance. The harmonized version of the Joint Committee of the European Centre for Disease Prevention and Control and the Centers for Disease Control and Prevention, which defined multidrug resistance as acquired non-susceptibility to at least one agent in three or more antimicrobial categories, was adopted for this study (Magiorakos *et al.*, 2012).

The antibacterial drugs used in an area provide suitable conditions for the emergence of multidrug resistant bacteria. All farms sampled in this study consulted the Veterinary Teaching Hospital of Ahmadu Bello University, Zaria at one point or another during the year 2014. It is however, a common practice among poultry farmers particularly small scale farmers and live bird vendors to seek veterinary advice after several failed attempts with several antibiotics. Melton (2014) showed that 54% of drugs used by poultry vendors were obtained without prescription from open markets and 68% of these vendors used their experience in determining drug dosages. Also, cocktails of antibiotics of different classes in a single formulation are commonly available in Nigerian markets. Therefore, poultry are frequently exposed to different classes of antibiotics at the same time thereby priming their microbial flora for multidrug resistance. This could explain why only one isolate was found to be sensitive to all antibiotics tested, while five others were sensitive either to only one or two antibiotics. This finding also almost correlates with that of Jahantigh and Dizaji (2015) in Iran; a country where poor drug use conditions are also said to prevail, who found that no *E. coli* isolate was sensitive to all antibiotics tested.

Fifty three (53) different resistance patterns ranging from resistance to a single class to a combination of seven (7) different classes of antibiotics were observed in this study. Olarinmoye *et al.* (2013) also reported thirty nine different resistance patterns with a multidrug resistance prevalence of 84.68% among E. coli in South Western Nigeria. This indicates the simultaneous involvement of various resistance mechanisms within some of these isolates. Multidrug resistance could be due to efflux pumps known to have broad substrate specificity (Hopkins *et al.*, 2005). For example, the AcrB protein of *E. coli*, in conjunction with other proteins such as TolC and AcrA, form a conduit which extrudes not only cellular metabolites but a diverse group of antibiotics (Soparkar *et al.*, 2015). In addition, certain plasmid-mediated resistance genes including *qnr*A and *qnr*B have been shown to be integrated into integrons which harbour other antibiotic resistance genes such as β-lactamases or aminoglycoside inactivating enzymes, while *qnr*S and *bla*TEM are known to be associated with transposons (Hernandez *et al.*, 2011). The involvement of mobile genetic elements such as plasmids, integrons and transposons in the spread of antibacterial resistance has emerged as a complicating factor in the effort to control resistance. These elements which are easily transmissible facilitate horizontal gene transfer thus promoting the evolution and adaptation of microbial communities (Smalla and Sobecky, 2002). Such developments serve to potentiate resistance in organisms with chromosomal resistance mechanisms to certain drugs such as fluoroquinolone resistance in Enterobacteriaceae.

All eighteen (18) isolates representing 24% of E. coli isolates subjected to PCR were shown to simultaneously carry qnrS and blaTEM genes. Although all these isolates were multidrug resistant, they were however, all sensitive to cefepime, ceftriaxone and imipenem with the exception of isolate no. 13 which was resistant to imipenem. Furthermore, while seventeen of these isolates were resistant to enrofloxacin, only five (5) were resistant to all FQs tested where as the other thirteen were resistant to other FQs at varying degrees. This pattern could result from the fact that cefepime, ceftriaxone and imipenem which are all β-lactams are not commonly used in poultry and even clinically they are reserved for the most resistant organisms whereas FQs are commonly used. Although *qnr* genes are known to be plasmid encoded, the absence of other *qnr* and ESBL genes in these isolates and the simultaneous occurrence of *qnr*S and *bla*TEM in all 18 isolates may suggest involvement of transposons or carriage of both genes on plasmids and transposons rather than integrons since *qnr*S is not known to be associated with integrons as reported by Hernandez *et al.* (2011). Both *qnr*S and *bla*TEM genes responsible for resistance to FQs and β-lactam antibiotics respectively, are very significant given that FQs are important first and second line drugs for treating clinically important infections (Thielman and Guerrant, 2004; Jamison, 2006), while certain β-lactam antibiotics such as carbapenem are drugs of last resort (Kunduru *et al.*, 2016). Fortini *et al.* (2011) reported the occurrence of genes involved in plasmid-mediated quinolone resistance (PMQR), *qnrS* inclusive and *blaTEM* in *E. coli* isolates from healthy chickens and pigs at slaughter in Ibadan, South Western Nigeria. Sumrall *et al.* (2014) also reported the mapping of *qnrS*1 gene from a Nigerian *E. coli* isolate to a 43.5 kb IncX2 plasmid (pEBG1). Variations in distribution of *qnr* genes has been suggested to be attributed to differences in geographic area (Oktem *et al.*, 2008)or selection criteria (Wang *et al.*, 2008). For example while *qnrS* was shown to be the most prevalent among members of *Enterobacteriaceae* family in studies in Egypt and China, the same gene was not detected in a US study (Robicsek, 2006; Wang *et al.*, 2008; El-Mahdy, 2015). Similarly, *qnrS* was identified as the most prevalent PMQR gene among *E. coli* isolates in a study involving wild birds and chickens in South Korea (Jae-Young *et al.*, 2016). The association of genes coding for resistance to FQs with those coding for resistance to other drug classes may facilitate selection and dissemination of FQ resistant strains by chemically unrelated drug classes, and vice versa (Dalhoff, 2012) thereby complicating the issue of antibacterial resistance. This suggests that resistance to a single drug may subsequently lead to resistance to many other drug classes.

5. CONCLUSION

This study has established the existence *E. coli* bearing *qnr*S and *bla*TEM from healthy chickens in Zaria, Northwestern Nigeria. As a readily available and relatively cheap source of protein to most Nigerians, healthy chickens and their products could serve as easy vehicle for transmission and spread of these resistance genes within the population with grave consequence for public health and the economy at large. Poultry farmers should therefore, use antibiotics as therapeutics only when necessary and in accordance with good drug use and management practices under the prescription of a certified veterinarian.

ACKNOWLEDGEMENTS

I am indeed most grateful to MacArthur foundation for sponsoring the MSc program on which led to the production of this paper. My sincere appreciation goes to Mal. Yahuza Maitala of Bacterial Zoonoses Laboratory, Department of Veterinary Public Health, Ahmadu Bello University (ABU), Zaria, and Mr. Jamok Jacob Elisha of the Centre for Biotechnology Research and Training (CBRT), ABU Zaria for their roles in the isolation and sensitivity testing and DNA extraction respectively. I will also like to acknowledge the Management of CBRT for granting me free access to and use of their facilities, kits and reagents for DNA extraction.

REFERENCES

[1] Aarestrup, F.M. (2000). Occurrence, selection and spread of resistance to antimicrobial agents used for growth promotion for food animals in Denmark. APMIS, 108(101):5–6. http://dx.doi.org/10.1111/j.1600-0463.2000.tb05380.x.

[2] Adenipekun, E.O., Jackson, C.R., Oluwadun, A., Iwalokun, B.A., Frye, J.G., Barrett, J.B., Hiott.

[3] Cattoir, V., Poirel, L. Rotimi, V, Soussy, C.J. and Nordmann, P. (2007). Multiplex PCR for detection of plasmid-mediated quinolone resistance *qnr* genes in ESBL-producing enterobacterial isolates. *Journal of Antimicrobial Chemotherapy,* 60(2):394–7. doi:10.1093/jac/dkm204

[4] Cohen, M.L. and Tauxe, R.V. (1986). Drug-Resistant *Salmonella* in the United States: An Epidemiologic Perspective. Science, 234(4779): 964–969. doi: 10.1126/science.3535069

[5] Dalhoff, A. (2012). Global Fluoroquinolone resistance Epidemiology and Implications for Clinical Use. Interdisciplinary Perspectives on Infectious Diseases. Article ID 976273 doi:10.1155/2012/976273.

[6] De Leener, E. (2005). Comparison of antimicrobial resistance among human and animal enterococci with emphasis on the macrolide-lincosamide-streptogramin group. Gent: UGent. https://lib.ugent.be/catalog/rug01:002151686 (accessed 15 Aug. 2014).

[7] El-Mahdy, R.H. (2015). Plasmid-Mediated Quinolone Resistance (PMQR) Determinants in Nosocomial Isolates of *Enterobacteriaceae. Egyptian Journal of Medical Microbiology,* 24(1): 43–47.

[8] Food and Agriculture Organization of the United Nations (2008). Joint FAO/WHO/OIE Expert Meeting on Critically Important Antimicrobials FAO Headquarters, Rome 26–30 November, 2007.

[9] Fortini, D., Fashae, K., Garcia-Fernandez, A., Villa, L. and Carattoli, A. (2011). Plasmid-mediated quinolone resistance and β-lactamases in *Escherichia coli* from healthy animals from Nigeria. *Journal of Antimicrobial Chemotherapy,* 66(6):1269–72.doi:10.1093/jac/dkr085

[10] Hernandez, A., Sanchez, M.B. and Martinez, J.L. (2011). Quinolone resistance: Much more thanpredicted. *Frontiers in Microbiology,* 2:22. doi: 10.3389/fmicb.2011.00022.

[11] Hopkins, K.L., Davies, R.H. and Threlfall, E.J. (2005). Mechanisms of quinolone resistance in *Escherichia coli* and *Salmonella*: recent developments. *International Journal of Antimicrobial Agents,* 25:358–373.doi:10.1016/j.ijantimicag, 2005.02.006.

[12] Jae-Young, O., Yong-Kuk, K., Dorji, T.M., Hyung-Kwan, J., Ok-Mi, J., Hee-Soo, L. and Min-Su, K. (2016). Plasmid-Mediated Quinolone Resistance in *Escherichia coli* Isolates from Wild Birds and Chickens in South Korea. Microbial Drug Resistance 22(1): 69–79. doi:10.1089/mdr.2015.0090.

[13] Jahantigh, M. and Dizaji, R.E. (2015). Antimicrobial Drug Resistance Pattern of *Escherichia coli* Isolated from Chickens Farms with Colibacillosis Infection. *Open Journal of Medical Microbiology,* 5: 159-162.doi:10.4236/ojmm.2015.54019.

[14] Jamison, D.T. (2006). World Bank Disease Control Priorities Project: Disease Control priorities in developing Countries (2nd Ed). Oxford University Press, Washington DC, USA.

[15] Johnson, J.R., Sanne, M.R., Johnston, B., Clabots, C., Kuskowski, M.A., Bender, J., Smith, K.E., Winokur, P.L. and Belongia, E.A. (2007). Antimicrobial Drug-Resistant *Escherichia coli* from Humans and Poultry Products, Minnesota and Wisconsin, 2002–2004. Emerging Infectious Diseases, 13:6.doi:10.3201/eid1306.061576.

[16] Kunduru, B.R., Nair, S.A. and Rathinavelan, T. (2016). EK3D: an *E. coli* K antigen 3-imensional structure database. Nucleic Acids Research, 44(D1): D675-D681. doi:10.1093/nar/gkv1313.

[17] L.M. and Woodley, T.A. (2015). Prevalence and Antimicrobial Resistance in *Escherichia coli* from Food Animals in Lagos, Nigeria. *Microbial Drug Resistance,* 21(3): 358–365. doi: 10.1089/mdr.2014.0222

[18] Magiorakos, A.P., Srinivasan, A., Carey, R.B., Carmeli, Y., Falagas, M.E., Giske, C.G., Harbarth, S., Hindler, J.F., Kahlmeter, G., Olsson-liljequist, B., Paterson, D.L., Rice, L.B., Stelling, J., Struelens, M.J., Vatopoulos, A., Weber, J.T. and Monnet, D.L. (2012). Multidrug-resistant, extensively drug-resistant and pandrug-resistant bacteria: an international expert proposal for interim standard definitions for acquired resistance. Clinical

Microbiology and infection, 18: 268–281. doi:10.1111/j.1469-0691.2011.03570.x.

[19] Melton, M.S. (2014). Antimicrobial resistance of *Escherichia coli* in poultry, knowledge and practice about antimicrobial use by poultry vendors in Yola. MPH thesis, A.B.U., Zaria.

[20] Monstein, H.J., Ostholm-Balkhed, A., Nilsson, N.V., Nilsson, M., Dornbusch, K. and Nilsson, L.E. (2007). Multiplex PCR amplification assay for rapid detection of *bla*SHV, *bla*TEM and *blaCTX-M* genes in *Enterobactericeae.* APIMS, 115: 1400-8.doi:10.1111/j.1600-0463.2007.00722.x.

[21] Nsofor, C.A., Iroegbu, C.U., Call, D.R. and Davies, M.A. (2013). Genetic relatedness of drug resistant *E. coli* isolates of human and animal origin in Nigeria. *International Journal of Genetics and Molecular Biology*, 5(3): 37–41. doi: 10.5897/IJGMB2013.0077.

[22] Ogunleye, A.O., Oyekunle, M.A. and Sonibare, A.O. (2008). Multidrug resistant *Escherichiacoli* isolates of poultry origin in Abeokuta, South Western Nigeria. VETERINARSKI ARHIV, 78(6): 501–509.

[23] Oktem, I.M., Gulay, Z., Bicmen, M. and Gur, D. (2008). *qnr*A prevalence in extended-spectrum betalactamase-positive *Enterobacteriaceae* isolates from Turkey. *Japanese Journal of Infectious Diseases*, 61(1): 13–7.

[24] Olarinmoye, A.O., Oladele, O.O., Adediji, A.A., Ntiwunka, U.G. and Tayo, G.O. (2013). Antibiograms of avian pathogenic Escherichia coli from commercial layers with colibacillosis in Southwest Nigeria. *Malaysian Journal of Microbiology*, 9(4): 317–325.

[25] Page, M.G.P. (2012). Antibiotic Discovery and Development. Springer, Boston, MA, USA, pp. 79. Doi: 10.1007/978-1-4614-1400-1_3.

[26] Paolo Pagani, J.E., Yerima, A. and Winnie, E. (2008). Assessment of the Nigerianpoultry market chain to improve biosecurity. *Food and agriculture organization of the united nations.*

[27] Paulo Martins da Costa, Luis, L. and Augusto, J.F.M. (2013). Transfer of Multidrug-Resistant Bacteria between Intermingled Ecological Niches: The Interface between Humans, Animals and the Environment. *International Journal of Environmental Research and Public Health*, 10: 278–294. Doi: 10.3390/ijerph10010278.

[28] Randall, L.P., Ridley, A.M., Cooles, S.W., Sharma, M., Sayers, A.R. and Pumbwe, L. (2003). Prevalence of multiple antibiotic resistance in 443 *Campylobacter spp.* isolated from humans and animals. *Journal of Antimicrobial Chemotherapy*, 52: 507–510.doi:10.1093/jac/dkg379

[29] Robicsek, A., Strahilevitz, J., Sahm, D.F., Jacoby, G.A. and Hooper, D.C. (2006). *Qnr* Prevalence in Ceftazidime-Resistant *Enterobacteriaceae* Isolates from the United States. *Antimicrobial Agents chemotherapy*, 50(8): 2872–2874. doi:10.1128/AAC.01647-05.

[30] Smalla, K. and Sobecky, P.A. (2002). The prevalence and diversity of mobile genetic elements in bacterial communities of different environ-

mental habitats: insights gained from different methodological approaches. *FEMS Microbiology Ecology,* 42(2): 165–175. https://doi.org/10.1111/j.1574-6941.2002.tb01006.x.

[31] Soparkar, K., Kinana, A.D., Weeks, J.W., Morrison, K.D., Nikaido, H. and Misra, R. (2015). Reversal of the drug binding pocket defects of the AcrB multidrug efflux pump protein of *Escherichia coli.* Journal of Bacteriology, 197: 3255–3264. doi:10.1128/JB.00547-15.

[32] Strahilevitz, J., Jacoby, G.A., Hooper, D.C. and Robicsek, A. (2009). Plasmid-mediated quinolone resistance: A multifaceted threat. *Clinical Microbiology Reviews,* 22(4): 664–689. doi: 10.1128/CMR.00016-09.

[33] Suarez, C. and Gudiol, F. (2009). Beta-lactam Antibiotics. Enfermedades Infecciosasy Microbiologia Clinica, 27(2): 116–129. Doi:https://doi.org/10.1016/j.eimc.2008.12.001

[34] Sumrall, E.T., Gallo, E.B., Oladipo, A.A., Lamikanra, A. and Okeke, I.N. (2014). Dissemination of the Transmissible Quinolone-Resistance Gene *qnr S*1 by IncX Plasmids in Nigeria. PLoS ONE, 9(10): e110279. doi: 10.1371/journal.pone.0110279.

[35] Thielman, N.M. and Guerrant, R.L. (2004). Clinical Practice. Acute Infectious Diarrhea. *New England Journal of Medicine,* 350(1): 38–47. doi:10.1056/NEJMcp031534.

[36] Van den Bogaard, A.E. and Stobberingh, E.E. (2000). Epidemiology of resistance to antibiotics. Links between animals and humans. *International Journal of Antimicrobial Agents,* 14: 327–335.http://dx.doi.org/10.1016/S0924-8579(00)00145-X

[37] Van den Boogard, J., Semvua, H.H., Boeree, M.J., Aarnoutse, R.E. and Kibiki, G.S. (2010). Sale of fluoroquinolones in Northern Tanzania: a potential threat for fluoroquinolone use in tuberculosis treatment. *Journal of Antimicrobial Chemotherapy,* 65(1): 145–147.doi: 10.1093/jac/dkp413.

[38] Wang, A., Yang, Y., Lu, Q., Wang, Y., Chen, Y., Deng, L., Ding, H., Deng, Q., Wang. L. and Shen, X. (2008). Occurrence of *qnr*-positive clinical isolates in *Klebsiella pneumoniae* producing ESBL or AmpC-type betalactamase from five pediatric hospitals in China. *FEMS Microbiology Letters,* 283(1): 112-6.DOI:10.1111/j.1574-6968.2008.01163.x.

Laboratory-based Antimicrobial Resistance Surveillance from Blood Cultures: Report from Different Hospital Setting in Indonesia

Ida Parwati[1*], **Adhi Kristanto Sugianli**[1] **and Andaru Dasihdewi**[2]

[1]Department of Clinical Pathology and Laboratory Medicine,
Faculty of Medicine, Padjadjaran University,
Dr. Hasan Sadikin General Hospital, Bandung, Indonesia

[2]Department of Clinical Pathology and Laboratory Medicine,
Faculty of Medicine, Public Health and Nursing,
Gadjahmada University, Dr. Sardjito Hospital, Yogyakarta, Indonesia

E-mail: *idaparwati2008@gmail.com

ABSTRACT: Antimicrobial resistance (AMR) is a global health problem, in low- and middle-income countries (LMICs). The AMR surveillance helps to monitor susceptibility patterns of microorganisms to antimicrobial and trend; and for assessing the impact of interventions. The data were collected between January–May 2018, using annual laboratory aggregate data. The participant hospital was requested to upload data into pre-designed report in web-based services (www. forlabinfeksi.or.id). Inclusion criteria were (1) result of blood culture and antibiotic susceptibility test in suspected or clinical infection patients; (2) valid specimen collection (3) microorganisms are considered as causative pathogens; and (4) test method using manual and/or automatic method. Submitted annual aggregate data was grouped into three stratifications based on hospital setting: primary-hospital (type-C); secondary hospital (type-B); and tertiary hospital (type-A). Thirty one hospitals participated in this study. The total number of isolates reported was 15,302 isolates, consist of 4761 Gram-positive isolates (31.1%) and 10541 Gram-negative isolates (68.9%). High resistance to Meropenem were observed in Klebsiella pneumoniae, Acinetobacter baumannii and Pseudomonas aeruginosa either in tertiary and primary hospital. Coagulase-negative Staphylococci (CNS) was most of isolated Gram-positive bacteria in blood specimen in all hospital type, and were high resistance to Oxacillin (methicillin-resistance CNS). Our study had shown a high antimicrobial resistant, both in Gram-positive and in

Gram-negative bacteria from blood-culture. Surveillance can be used to describe the trend of circulating resistant pathogen, although several issues have been identified as limitation, particularly for laboratory-based surveillance. Further suggestion, AMR surveillance program should be implemented nationwide, to gain a representative national AMR situation.

Keywords: Antimicrobial resistance, Surveillance, Blood Culture, Gram-negative, Gram-positive.

1. INTRODUCTION

Bloodstream infection (BSI) is the most serious form of healthcare-associated infection and is associated with significant morbidity and mortality. By definition, an episode of BSI is defined as a positive blood culture from one or more blood sample and having one or more recognised bacterial or fungal pathogens.[1,2] Currently, the multidrug resistant bacteria, which manifest as bacteremia, leads as a burden anti-microbial resistance issue and presents therapeutic challenges in hospital.[2,3] The AMR problem is happened globally, in low- and middle-income countries (LMICs). The AMR in Indonesia is threatening since the prevalence of infectious disease is high and worsen by weak antibiotic policies. The lack of antibiotic resistance surveillance capacity, contribute to antibiotic resistance as well.[4,5] The WHO Global Strategy for Containment of Antibiotic Resistance (GLASS) recognizes laboratory-based surveillance of antibiotic resistance as a fundamental priority for the development of strategies to contain antibiotic resistance. The AMR surveillance helps to monitor susceptibility patterns of microorganisms to antimicrobial and periodically updated to see trend of pattern changes; and for assessing the impact of interventions. Furthermore, an efficient AMR surveillance system is necessary to help reducing mortality and morbidity.[6]

It has been indicated that very little data is available about AMR situation in Indonesia, mostly based on specific studies, which conducted by laboratories or universities.[4] Therefore, in this study, we would like to describe the current situation of AMR in Indonesia, particularly in blood culture by determining the prevalence of pathogenic bacteria and its susceptibility to certain antimicrobial drugs among different hospital settings.

2. MATERIAL AND METHODS

2.1 Surveillance System Model

Pre-surveillance meeting was conducted to have good understanding and perception about the surveillance system. The participants were from different hospital setting in Indonesia and were asked for their agreement to collect the aggregate data based on their routine laboratory work. The system model was performed according to the WHO-GLASS guideline.[6] This study was approved by ethical committee of Dr. Hasan Sadikin Hospital Bandung, No: LB.04.01/A05/EC/374/XII/2016.

2.2 Data Collection

The data collection was performed between January–May 2018, using annual laboratory aggregate data (time frame period January–December 2017). The participant hospitals were requested to upload data into pre-designed report in web-based services (www.forlabinfeksi.or.id). The inclusion criteria for aggregated data were:

1. Result of culture examination and antibiotic susceptibility test in suspected or clinical infection patients;
2. Valid specimen collection;
3. Microbial are considered as causative pathogens (based on comprehensive expertise); and
4. Test method using manual and/or automatic method, with identification probability criteria $\geq$ 80% and the result of antibiotic susceptibility test is consistent.[7]

Each of participating hospital was also monitored for the quality control of identification and susceptibility testing, according to the CLSI guideline.[8] Submitted annual aggregate data was grouped into three stratifications based on hospital setting: primary-hospital (type-C); secondary hospital (type-B); and tertiary hospital (type-A). Type of specimen from each annual aggregate data was grouped into blood, sputum, urine, stools, pus, cerebrospinal fluid (CSF), wound, and other body fluids. Meanwhile, for the source of specimens was further stratified into outpatient, inpatient, intensive care, and emergency. We did not use intensive care as part of surveillance, since intensive care could skew the antimicrobial susceptibility results due to severity of case and previous antimicrobial drugs usage. Therefore, we only used inpatient and blood specimen for data analysis as the objectives of this study.

2.3 Data Analysis

The aggregate data from participant hospital was analysed using WHONET 5.6 software (WHO Collaborating Centre for Surveillance of Antimicrobial Resistance, 2019). During data analysis of positive blood-culture, specific criteria should be met as described in Clinical Laboratory Standard Institute (CLSI) documentation-M39A.[9] Bacterial pathogen and antimicrobial susceptibility report were generated using the WHONET software and Microsoft Excel. Bacterial prevalence was reported in total number of isolate among all specimens (N); total number of bacterial being identified in blood specimens (n); and percentage of isolate. The bacterial percentage was defined as percentage (%) of bacterial from positive-blood specimen to total bacterial isolated from positive-culture of all specimens. Susceptible percentage was defined as percentage (%) of isolate that susceptible to certain antimicrobial among all isolates being tested. The percentage of antimicrobial resistance was calculated by subtraction of 100% with susceptible percentage.

3. RESULT

During data collection, there were 31 hospitals participating in this study. The participant included 22 public hospitals and 9 private hospitals, which consist of: 26 hospitals in west part (Java, Sumatera); 4 hospitals in central part (Borneo, Bali, Sulawesi) and 1 hospital in east part (Papua, Maluku) of Indonesia. The total number of isolates reported was 15,302 isolates, consist of 4761 Gram-positive isolates (31.1%) and 10541 Gram-negative isolates (68.9%).

The percentage of Gram-negative and Gram-positive bacteria isolated from blood cultures were shown in Table 1. The number of positive Gram-positive bacteria isolated from all specimen in secondary hospital was less than 30 isolates; therefore, the number of positive blood culture was not reported for this study. *Escherichia coli* isolate from secondary and primary hospital was also less than 30 isolates, but this organism still accounted for surveillance report due to their pathogenic characteristic in hospital. The positivity-rates of blood culture for

Gram-negative and positive bacteria were ranging between 5% to 34%, except for *coagulase-negative staphylococci* (CNS) (37–55%) and *Salmonella typhi* (100%). The prevalence of bacterial isolates from blood culture in hospital setting were dominated by CNS. Gram-negative bacteria as pathogenic organism was identified in blood culture, dominated by *Acinetobacter baumannii* (10-13%), *Klebsiella pneumoniae* (6–11%) and *Escherichia coli* (6–7%). *Salmonella typhi* was the only Gram-negative bacteria that isolated in secondary hospital setting.

Table 1: The Percentage of Blood Pathogen from Culture-Positive among Different Hospital Settings

Hospital Settings	*Organism*	*Positive Culture of All Specimen*	*Positive Blood Culture*	*%*
Tertiary Type A	*Escherichia coli*	1302	89	7%
	Klebsiella pneumoniae	1100	99	9%
	CNS	721	267	37%
	Pseudomonas aeruginosa	660	33	5%
	Acinetobacter baumannii	648	62	10%
	Staphylococcus aureus	283	68	24%
Secondary Type B	*Escherichia coli*	414	24	6%
	Klebsiella pneumoniae	295	37	13%
	Salmonella typhi	48	48	100%
Primary Type C	*Klebsiella pneumoniae*	589	67	11%
	Acinetobacter baumannii	506	67	13%
	Escherichia coli	447	28	6%
	Staphylococcus aureus	199	41	21%
	CNS	553	304	55%

Abbrev: %, bacterial percentage; CNS, coagulase-negative staphylococci.

Table 2 showed the surveillance result of antimicrobial resistance in Gram-negative bacteria from positive blood culture, stratified by different hospital settings. High antimicrobial resistance was observed in *Klebsiella pneumoniae, Acinetobacter baumannii* and *Pseudomonas aeruginosa*. Tigecycline, as preserve antimicrobial drugs, showed low resistance among all of Gram-negative bacteria, ranging from 0–26%. High resistances to Meropenem were observed in *Klebsiella pneumoniae, Acinetobacter baumannii* and *Pseudomonas aeruginosa,* except for *Escherichia coli* and *Salmonella typhi*. *Salmonella typhi* was the only organism that observed in secondary hospital and showed low resistance to most of antimicrobial drugs, except co-trimoxazole and aminoglycoside group (gentamicin and amikacin).

Table 2: Antimicrobial Resistance of Gram-negative Bacteria, Isolated from Blood Culture in Different Hospital Settings

Antimicrobial Drugs	*K. pneumoniae*			*E. coli*			*A. baumannii*		*P. aeruginosa*	*S. typhi*
	Type A	*Type B*	*Type C*	*Type A*	*Type B*	*Type C*	*Type A*	*Type C*	*Type A*	*Type B*
	n = 99	*n = 37*	*n = 67*	*n = 89*	*n = 24*	*n = 28*	*n = 62*	*n = 67*	*n = 33*	*n = 48*
	%R	*%R*	*%R*	*%R*	*%R*	*%R*	*%R*	*%R*	*%R*	*%R*
GEN	72	80	73	34	12	19	73	78	46	100
AMK	42	3	28	1	0	0	64	54	25	100
SAM	85	83	91	69	59	71	68	70	67	0
FEP	90	80	84	46	29	42	81	80	46	0
CAZ	88	80	ND	67	29	ND	85	ND	33	0
CRO	88	80	84	69	29	46	98	94	100	0
CIP	35	74	56	70	33	40	81	79	38	0
LVX	48	ND	25	75	ND	75	79	0	50	ND
SXT	63	80	59	66	47	40	32	58	100	100
TZP	53	74	60	16	11	10	83	89	33	0
TGC	3	6	26	0	0	13	10	24	0	0
MEM	11	51	34	4	0	4	57	73	28	0

Abbrev: GEN, gentamicin; AMK, amikacin; SAM, ampicillin-sulbactam; FEP, cefepime; CAZ, ceftazidime; CRO, ceftriaxone; CIP, ciprofloxacin; LVX, levofloxacin; SXT, co-trimoxazole; TZP, piperacillin-tazobactam; TGC, tigecycline; MEM, meropenem; n, number of isolate being identified and tested for susceptibility testing.

Table 3 showed the surveillance result of antimicrobial resistance in Gram-positive bacteria from positive blood culture, stratified by different hospital settings. *Coagulase-negative Staphylococci,* such as *Staphylococcus hominis, Staphylococcus epidermidis* and *Staphylococcus haemolyticus,* was most of isolated bacteria in blood specimen among all the hospital settings, which generally classified as contaminant bacterial.[2] High resistance to Oxacillin were observed among the CNS group bacteria, which indicated resistance to methicillin (methicillin-resistance CNS). Compared with CNS group bacteria, resistance to

Oxacillin among *Staphylococcus aureus* was observed lower (27–31% vs 74–94%); this indicates the presence of methicillin-resistance *Staphylococcus aureus* (MRSA) was low in the hospital setting. Vancomycin, as preserved antibiotic for MRSA and methicillin-resistance CNS treatment, showed low resistance, except in primary hospital (type C) for *Staphylococcus aureus.*

Table 3: Antimicrobial Resistance of Gram-positive Bacteria, Isolated from Blood Culture in Different Hospital Settings

Antimicrobial Drugs	*CNS*						*S. aureus*	
	S. hominis		*S. epidermidis*		*S. haemolyticus*			
	Type A	*Type C*	*Type A*	*Type C*	*Type A*	*Type C*	*Type A*	*Type C*
	n = 97	*n = 109*	*n = 94*	*n = 59*	*n = 76*	*n = 136*	*n = 68*	*n = 41*
	%R	*%R*	*%R*	*%R*	*%R*	*%R*	*%R*	*%R*
GEN	28	4	76	42	73	70	16	23
VAN	4	12	7	0	7	4	9	33
CIP	60	34	57	68	88	80	21	27
LVX	57	33	60	69	87	76	10	31
LNZ	2	6	0	3	3	4	4	0
ERY	74	71	84	80	84	79	23	31
MFX	52	ND	46	ND	71	ND	14	ND
CRO	84	73	91	92	91	93	26	32
OXA	84	74	90	92	91	94	27	31
TGC	1	10	0	20	2	33	0	40
SXT	77	49	62	60	29	50	9	55
CLR	84	69	81	75	74	81	16	23

Abbrev: CNS, *coagulase-negative Staphylococci*; GEN, gentamicin; VAN, vancomycin; CIP, ciprofloxacin; LVX, levofloxacin; LNZ, linezolid; ERY, erythromycin; MFX, moxifloxacin; CRO, ceftriaxone; OXA, oxacillin; TGC, tigecycline; SXT, co-trimoxazole; CLR, clarithromycin; ND, not detected; n, number of isolate being identified and tested for susceptibility testing.

We also performed observation between the surveillance result and the Indonesian Drugs Formulary[10] for selected isolate and antimicrobial drugs, as shown in Figure 1. The antimicrobial drugs that commonly prescribed, such as fluoroquinolone, co-trimoxazole and cephalosporin, were observed having high resistance for *Escherichia coli, Klebsiella pneumoniae* and *Staphylococcus aureus,* in all hospital settings. Interestingly, co-trimoxazole and levofloxacin showed low resistant against *Staphylococcus aureus* in tertiary hospital, compared with primary hospital.

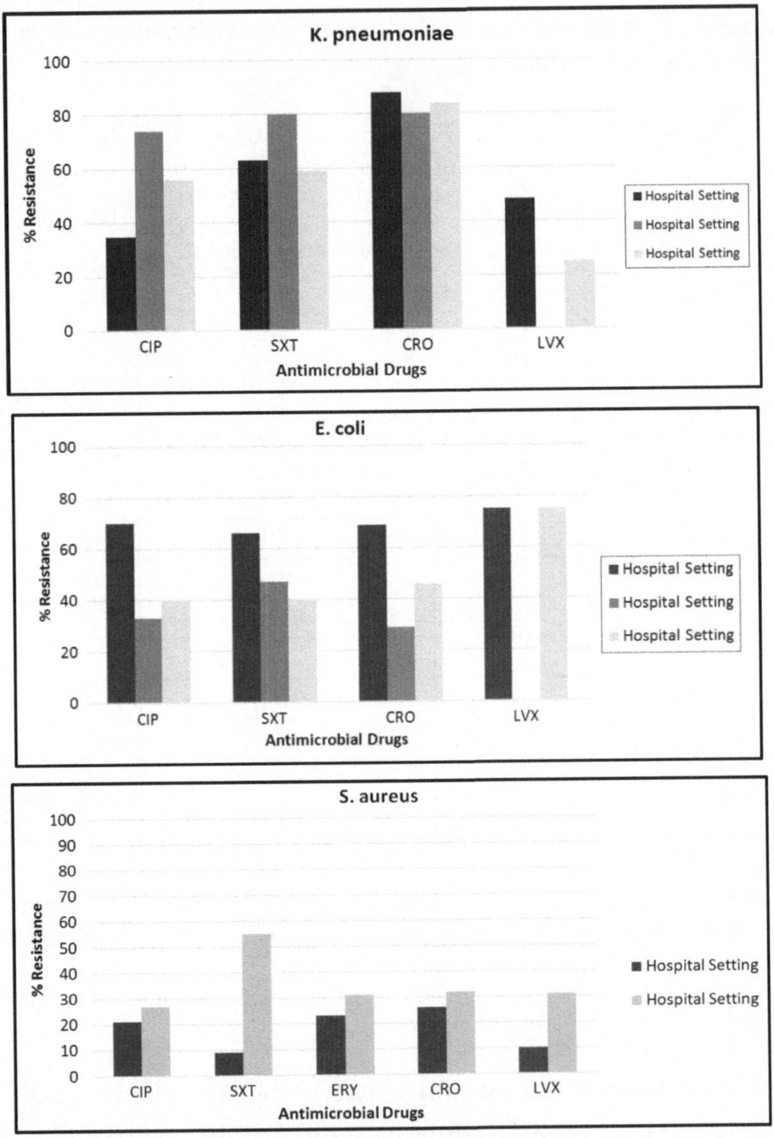

Figure 1: Percentage of Resistance of *Klebsiella pneumoniae, Escherichia coli* and *Staphylococcus aureus* to Recommended Antimicrobial Drugs in Hospital, Stratified by Hospital Settings.
Abbrev: CIP, ciprofloxacin; SXT, co-trimoxazole; CRO, ceftriaxone; LVX, levofloxacin; ERY, erythromycin.

4. DISCUSSION

This is the first reported study of AMR in Indonesia involving both public and private hospital setting, which represent West-, Central-, and East-Indonesia area. The prevalence of AMR to Gram-negative bacteria in blood culture was dominantly by *Klebsiella pneumoniae* and *Acinetobacter baumannii* isolates. These isolates are mostly found in the bloodstream infections, and has been known widely contribute to AMR profile, such as 3rd generation cephalosporin-resistance and carbapenem-resistance.[1,2] Data report from WHO in South East Asia also find similar results whereas resistance to 3rd generation cephalosporins, reaching 34–81%.[11] Although this study was a laboratory-based surveillance, but this information contributes to nation problem, since the data are reflecting from different hospital settings and alarmingly high rates of resistances in most of identified pathogens.[11,12] Interestingly, in Gram-positive bacteria isolated from positive-blood culture, only two group species were found dominantly: *Staphylococcus aureus* and CNS. Moreover, the prevalence of MRSA was lower than methicillin-resistant CNS in positive blood culture. CNS bacteria are the most common bacteria cause nosocomial BSI and related with healthcare-associated bacteremia.[2] All blood-positive specimens were taken for this study from inpatient care, excluding the intensive care. Therefore, the data shows similarity with previous reported work.[2,12] It has been reported that resistance rates for CNS up to 90% for methicillin, 78.6% for levofloxacin, 68% for ciprofloxacin, and 48.5% for clindamycin in US between 2014–2016.[2] No vancomycin-resistance or vancomycin-intermediate *Staphylococcus aureus* was observed in this study for different hospital settings.

According to our finding, we agree with previous study that AMR surveillance data can reflect an increasing trend of resistance of important bacterial pathogens, either community or hospital transmitted infections. As reported by WHO, infrastructure for laboratory based surveillance was observed poor. Nationally coordinated laboratory services become a low priority in most South-east Asian countries, with minimal funding. Furthermore, several issues are also identified as limitation of surveillance, such as (1) no agreement on surveillance standards, resulting in a lack of reliable and comparable data; (2) most of surveillance reporting the proportions of resistant bacteria causing specific diseases or affecting defined populations.

Coordination between laboratories, hospital administration, and policy makers is also insufficient to collect a representative data.[11,12]

We observed limitation of this study. Previous study had observed potential bias induced by laboratory-based surveillance, which might be skewed toward higher prevalence of AMR.[13–15] Since we used aggregate data from routine laboratory work, the potential biases were selection processes for submission of clinical specimens to laboratories. Different guidelines among physicians would interfere their clinical judgement to submit any specimen for culture and susceptibility testing.

5. CONCLUSION

Our study shows a high antimicrobial resistant, both in Gram-positive and in Gram-negative bacteria from blood-culture. Surveillance can be used to describe the trend of circulating resistant pathogen, although several issues have been identified as limitations, particularly for laboratory-based surveillance. Further suggestion, AMR surveillance program should be implemented nationwide, to gain a representative data of national AMR situation.

ACKNOWLEDGMENTS

I.P., A.K.S., A.D. contributed equally to this work. I.P. conceptualized and drew up the first manuscript; A.K.S did the literature review and wrote the manuscript. A.K.S, A.D. did the data collection. I.P., A.K.S., A.D. read and finalized the final manuscript. We would like to thank our colleagues for their contribution of the surveillance data. This work is part of our contribution to the Indonesian Association of Clinical Pathology and Laboratory Medicine. No funding and conflict of interest was declared.

REFERENCES

[1] Bharadwaj, R., Bal, A., Kapila, K., Mave, V. and Gupta, A., Blood Stream Infections. *BioMed Res Int.*, 2014; 2014: 1–3.

[2] Akova, M., Epidemiology of antimicrobial resistance in bloodstream infections. *Virulence,* 2016 Apr 2; 7(3): 252–66.

[3] Viscoli, C. Bloodstream Infections: The peak of the iceberg. Virulence, 2016 Apr. 2; 7(3): 248–51.

[4] Indonesian Ministry of Health. National Action Plan on Antimicrobial Resistance Indonesia [Internet]. Ministry of Health; 2017. Available from:

http://apps.who.int/datacol/answer_upload.asp?survey_id=666&view_id=722&question_id=13163&answer_id=19958&respondent_id=227949

[5] World Health Organization. Global Action Plan on Antimicrobial Resistance [Internet]. World Health Organization, 2015. Available from: http://www.who.int/iris/bitstream/10665/193736/1/9789241509763_eng.pdf?ua=1

[6] World Health Organization. Global Antimicrobial Resistance Surveillance System: Manual for Early Implementation [Internet]. World Health Organization; 2015. Available from: https://apps.who.int/iris/bitstream/handle/10665/188783/9789241549400_eng.pdf?sequence=1

[7] Dahesihdewi, A., Sugianli, A.K. and Parwati, I., The surveillance of antibiotics resistance in Indonesia: a current reports. *Bali Med J.*, 2019; 8(2):474–9.

[8] Clinical and Laboratory Standards Institute. Performance Standards for Antimicrobial Disk Susceptibility Tests - Eleventh Edition: Approved Standard M02-A11. Wayne, Pennsylvania: Clinical and Laboratory Standards Institute; 2012.

[9] Clinical Laboratory Standard Institute. Analysis and Presentation of Cumulative Antimicrobial Susceptibility Test Data; approved guideline - Third Edition M39-A3. Wayne, Pennsylvania: Clinical and Laboratory Standards Institute; 2009.

[10] Indonesian Ministry of Health. Formularium Nasional Nomor HK.01.07/MENKES/659/2017 [Internet]. Indonesian Ministry of Health; 2017. Available from: http://hukor.kemkes.go.id/uploads/produk_hukum/KMK_No._HK_.01_.07-MENKES-659-2017_ttg_Formularium_Nasional_.pdf

[11] World Health Organization Regional Office for South-East Asia. Situation analysis on antimicrobial resistance in the South-East Asia Region: Report 2016 [Internet]. World health organization, Regional Office for south-east Asia; 2017. Available from: https://apps.who.int/iris/handle/10665/272873

[12] Shah AS, Karunaratne K, Shakya G, Barreto I, Khare S, Paveenkittiporn W, *et al.* Strengthening laboratory surveillance of antimicrobial resistance in South East Asia. BMJ. 2017 Sep 5; j3474.

[13] Rempel, O.R. and Laupland, K.B. Surveillance for antimicrobial resistant organisms: potential sources and magnitude of bias. Epidemiol Infect. 2009 Dec.; 137(12):1665.

[14] Rempel, O., Pitout, J.D.D. and Laupland, K.B. Antimicrobial Resistance Surveillance Systems: Are Potential Biases Taken into Account? *Can J Infect Dis Med Microbiol.* 2011; 22(4): e24–8.

[15] Laupland, K.B., Ross, T., Pitout, J.D.D., Church, D.L. and Gregson, D.B. Investigation of sources of potential bias in laboratory surveillance for anti-microbial resistance. *Clin Invest Med.* 2007 Aug 1; 30(4):159.

Antibiogram Profile of Selected Clinical Microbial Isolates from University of Benin Teaching Hospital (UBTH), Benin City, Nigeria

Dowe Ejiro[1*], Ahonkhai Ijeoma[2] and Odiete Congrat Eravweroso[3]

Department of Pharmaceutical Microbiology Faculty of Pharmacy, University of Benin, Benin City, Edo State, Nigeria

E-mail: *ejiro.dowe@pharm.uniben.edu

ABSTRACT: The development of resistance to antimicrobial agents by clinical isolates is becoming alarming, due to the increase in incidence of nosocomial infections and frequent exposure to conventional antibiotics. The antibiogram profile of selected antibiotics against clinical bacterial and fungal isolates in relation to their standard type strains were investigated using the Agar disc diffusion and broth dilution method. The results of the Inhibition Zone Diameters (IZDs) and Minimum Inhibitory Concentration (MIC) shows that all the clinical isolates were relatively more resistant or rather, less susceptible compared to their standard type strains. Bacillus subtilis were more susceptible to the antibiotics, followed by, Streptococcus pyogenes, Proteus vulgaris, Escherichia coli, Klebsiella pneumoniae, Staphyococcus aureus and Pseudomonas aeruginosa. Among the fungal isolates, Cryptococcus neoformans was more susceptible than Candida albican with an interpreted resistance IZD to the inhibitory effect of fluconazole. E. coli was resistant to the inhibitory effect of erythromycin and doxycycline. In addition, K. pneumoniae was resistant to erythromycin, trimethoprim and doxycycline. Multidrug resistance was recorded for S. aureus to the inhibitory effect of co-amoxiclav, erythromycin, gentamycin, trimethoprim and doxycycline while P. aeruginosa also recorded Multi-Dug Resistance (MDR) to the inhibitory effect of co-amoxiclav, erythromycin, cefoxitin, trimethoprim and doxycycline. The finding from this study validates the trend of emerging resistance and multi-dug resistance clinical isolates that has become a global health problem.

Keywords: Antibiogram, Antibiotics, Clinical, Isolates, MIC, IZD, Sensitivity, Susceptible.

1. INTRODUCTION

Clinical microbial isolates are microorganisms isolated from either of several cases of patients that visit healthcare centers due to varying ailments (Prescot *et al.,* 2005, Kreth *et al.,* 2009, Dowe *et al.,* 2016). According to Bachman *et al.* (2017), microorganisms or infectious agents such as bacteria, fungi, viruses and parasites cause various form of ailments and diseases. In most clinical or healthcare setting, infection diagnosis and treatment begins with sample collection from patients at a body site or fluid where the detection of pathogens or its associated biomarkers is likely to signify disease(s). Obtained samples is immediately transported to the microbiological or diagnostic laboratory where they are analyzed or tested in a way that is sensitive and specific for the suspected causative agent of the disease using standard microbiological methods (Usman *et al.,* 2005, Schwendicke *et al.,* 2015). Disease causing microorganisms may be isolated from oral swabs (odonthopathogens), urine samples, vaginal swabs, body surfaces, wounds, other body fluids, etc. The development of resistance to antimicrobial agents by clinical isolates is becoming alarming, due to the increase in incidence of nosocomial infections and increasing microbial resistance to conventional antimicrobial agents (WHO 2010).

Antimicrobials such as antibiotics, antiplasmodial, antifungal and antiviral agents inhibit or destroys microbial cells at their effective concentrations or standard doses and have equally played key role in prevention and treatment of diseases or infectious agents (Salim *et al.,* 2005; WHO 2010). However, together with the use of antimicrobial medicines are issues of resistance. According to Salim *et al.* (2005), the discovery of the first antibiotic known as penicillin by Alexander Fleming in 1928, retarded the prevalence of infectious diseases and saved millions of lives. In his Nobel price speech in 1945, Fleming stated that bacteria could develop resistance against antibacterial therapies and it was not much later when cases of non-efficacy of the "wonder drug" itself were reported. This started the reporting of failure of other therapeutic drugs against infectious agents, later recognized as drug resistance, antimicrobial resistance (AMR) or antibiotic resistance challenging the efficacy of modern therapeutic regimes. AMR has become a global health problem, currently considered as an emergent disease and a major public health problem (WHO 2014). Contributing to the existing problem is the spread of cross-resistance and multidrug resistance (MDR) genes among pathogenic organisms. Infectious agents of all sorts including those of viral, bacterial fungal protozoans, etc. origin has evolved various mechanisms of circumventting or resisting chemotherapeutic agents (John and James 1999, Dowe *et al.,* 2016).

The emergence of resistant microorganisms either the emergence of resistant microorganisms either by mutation or the acquisition of mobile genetic elements bearing resistant genes has been accelerated by frequent exposure to antimicrobial agents such as antibiotics which provides the necessary selective pressure for the rise and spread of resistant pathogens. Similarly, inappropriate and irrational use of antibiotics in man and animals is a major factor that promotes the spread of resistance. The aim of this study is to investigate the antibiogram profile of selected antibiotics against clinical bacterial and fungal isolates in comparison to their standard type strains.

2. MATERIALS AND METHODS

2.1 Materials

Materials used include table-sized autoclave, mettler-weighing balance and glasswares of pyrex, England. All antibiotics disc and powders were of Oxoid Ltd., England. Microbiological media were obtained from Biotech., India and include Mueller Hinton agar, Nutrient broth, Sabouraud Dextrose agar and Sabouraud Dextrose broth.

2.2 Study Site

The site of sample collection was the University of Benin Teaching Hospital (UBTH) located on the same geographical coordinates with the university of Benin (UNIBEN); latitude 6°20′ 1.32″ N and longitude 5°36′ 0.53″ E. Stock cultures of microbial isolates from oral infections, wound swabs and urine samples were sourced from the

Medical Microbiology department of UBTH. Isolates were authenticated and used for the study.

Figure 1: Map of Nigeria Showing Benin City where UNIBEN/UBTH is Located (*Source:* Google maps 2018)

2.3 Sources of Test Organisms/Bacterial Culture

The bacterial cultures used for this study were obtain from stock cultures of isolates characterized and authenticated from the treatment of oral swabs, wound swabs and urine samples collected from UBTH, Benin City, Edo state, Nigeria. On the other hand, the standard bacterial cultures were from stock of American type culture collection (ATCC) United States of America. All the standard typed cultures used were obtained from previously authenticated stock cultures preserved in the Department of Pharmaceutical Microbiology and Biotechnology Laboratory, Faculty of Pharmacy, University of Benin, Nigeria. The clinical isolates used in this study were: *Staphylococcus aureus*, *Streptococcus pyogenes*, *Escherichia coli*, *Klebsiella pneumoniae*, *Pseudomonas aeruginosa*, *Bacillus subtilis*, *Proteus vulgaris* and *Cryptococcus neoformans*, *Candida albicans* while the standard strains used were;

Escherichia coli ATCC 25922, *Staphylococus aureus* ATCC 25923, *Streptococcus pyogenes* 12384, *Bacillus subtilis* ATCC 21332, *Proteus vulgaris* 29905, *Klebsiella pneumoniae* 2342, *Pseudomonas aeruginosa* 29260, *Cryptococcus neoformans* ATCC 90112 and *Candida albicans* 10259.

2.4 Antimicrobial Sensitivity to Selected Antibiotics

Antibiotic susceptibility of the isolates was tested by the modified Kirby-Bauer disk diffusion method described by Lalitha *et al.* (2004). Seven commercially prepared oxoid antibiotic disc (co-amoxiclav 30 µg, Erythromycin 15 µg, Gentamycin120 µg, Cefoxitin 30 µg, Doxycycline 30 µg, Trimethoprim 5 µg and Ciprofloxacin 5 µg) and three commercially preparedoxoid antifungal disc (Clotrimazole 15 µg, Fluconazole 30 µg and Ketokonazole 30 µg) were placed on the surface of seeded agar (Mueller Hinton for bacterial and SDA for fungi) plates using a flamed forceps. Prior to seeding, the turbidity of the overnight broth culture of each isolates were adjusted to match 0.5 McFarland turbidity standard and was further diluted one in hundred (1:100) serial dilution to give approximately 1×10^6 cfu/ml microbial suspension. Sterile swab sticks were then dipped into the standardized microbial suspension and gently rolled against the wall of the test-tube to remove excess fluid and the swab stick was streaked on the surface of the agar plates in even strokes to obtain a uniform growth pattern across the entire surface of the plate. This was achieved by rotating the plate 90 degrees followed by 45 degrees with continuous streaking, and finally by streaking round the diameter of the agar. The agar plates were allowed to absorb the fluid and the antibiotic discs were then placed on the seeded agar surface using flamed forceps. Lids were replaced and plates appropriately incubated. All experiments done in triplicates, plates observed for presence or absence of clear zones, and the inhibition zone diameters measured.

2.5 Determination of Minimum Inhibitory Concentration (MIC) of the Drugs

The modified broth dilution method described by Lalitha *et al.* (2004). Varying concentrations of the drug ranging from 0.01–10 µg/ml were constituted in 10 ml of Mueller-Hinton broth in sterile capped tubes from the already made stock and a drop of the standardized microbial suspension of each organism was added to the tube using 0.025 ml dropper pipette. Overnight broth culture of the test organisms was standardized to 0.5 Mc Farland turbidity standard corresponding to approximately 1×10^8 cfu/ml and was further diluted (1:100), corresponding to microbial suspension of approximately 1×10^6 cfu/ml. The same experimental set-up using the standard type culture served as positive control and tube without the drug but with same volume of broth served as negative controls. The same experiment was repeated for the fungal organisms but Sabouraud dextrose broth was used in place of Mueller-Hinton. All tubes were appropriately incubated. After incubation, tubes were observed for growth by examining for turbidity and plating out the content of target tubes on agar plates, followed by incubation of bacterial plates at 38°C for 24 hours and fungal plates at 28 ± 2°C for 48 hours. In all cases, the lowest concentration of the antibiotics at which there was no observable bacterial or fungal growth was recorded as the MICs.

3. RESULTS AND DISCUSSIONS

The results of the antibiogram profile of the selected microbial isolates shows the least set of inhibition zone diameters (IZDs) for Erythromycin (E), followed by Doxyclines (DOX), and Trimethoprim (W) while Co-amoxiclav (AMC) and Gentamycin (GN) recorded intermediate IZDs relative to the two extremes, whereas Cefoxitin (FOX) and Ciprofloxacin (CIP) recorded the highest IZDs (see Figure 2). This corroborates the study reported by Doern (1995) who recorded lower set of IZDs for Penicillins, followed by the tetracyclines and further recorded the highest set of IZDs for the quinolones. The antimicrobial activity of antimicrobial agents was determined by the presence of IZD on seeded agar plate (Cheesbrough, 2006). A stronger antimicrobial agent will create a larger IZD or inhibit the growth of microbial cells at a relatively lower concentration (Cheesbrough, 2006, Dowe *et al.*, 2016). According to Doern (1995) the quinolones antibiotics such as ciprofloxacin is a stronger antibiotic compared to lower antibiotics such as tetracyclines. Their derivatives and production was in response to the problem of emerging resistant strains of bacterial isolates to previously existing antibiotics. In addition, among the antifungal antibiotics, clotrimazole (15 μg) recorded the highest IZDs, followed by Ketoconazole (30 μg) and the least was fluconazole (30 μg) to which *C. albican* was resistant. These antifungal agents belong to the same azole group and are known to exert their actions by disrupting fungal membrane sterols or more specifically ergosterol by inhibiting the enzyme cytochrome P_{450} dependendent 14-α-demethylation of lanosterol (Andrew 2001).

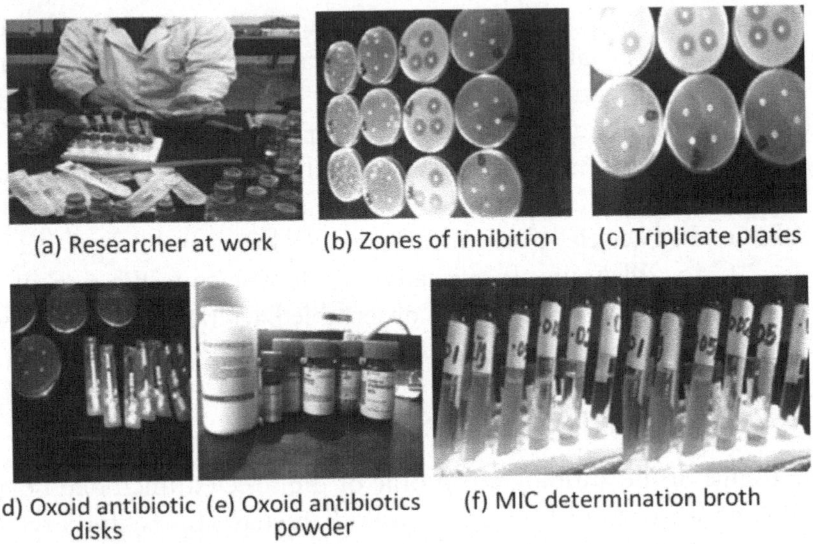

Figure 2: (**a**) Researcher at Work, (**b–c**) Zones of Inhibition on Plates, (**d–e**) Antimicrobial Agents and (**f**) MIC Determination Broths

Similar pattern of IZD was recorded for the standard type culture against the various antibiotics. However, the IZD results were higher for the standard type strains as against those recorded for the isolates which were in some cases resistant to the inhibitory effect of the antibiotics. The interpretation of the results of IZDs (see Table 2), recorded higher resistance for *P. aeruginosa* and *S. aureus* isolates. *Staphylococcus aureus* was resistant to co-amoxiclav (30 µg), erythromycin (15 µg), gentamycin (120 µg), trimethoprim (5 µg) and doxycycline (30 µg) wher as *P. aeruginosa* isolates recorded resistance to the inhibitory effect of co-amoxiclav (30 µg), erythromycin (15 µg), cefoxitin (30 µg), trimethoprim (5 µg) and doxycycline (30 µg). However, none of the standard type test organisms were resistant even though some did show intermediate susceptibility, a large percentage were interpreted as susceptible. In general, comparing the IZDs of the isolates to the standard type culture, some level of resistance could be traced due to the large contrast. According to Lalitha *et al.* (2004), interpreted IZDs seem to permit liberal fluctuations to accommodate some limit of errors and in some cases, suggest intermediate susceptibility when in reality results should be strictly interpreted as resistance, hence enzymatic assay seem to remove such limitations. Organisms with 10 µg Penicillin disc zone diameters of < 19 mm generally produce ß-lactamase. However, ß-lactamase tests are faster and are therefore preferred for recognition of this plasmid-mediated resistance. Organisms with plasmid-mediated resistance to tetracycline also have zones of inhibition (30 µg tetracycline discs) of < 19 mm. Chromosomal mechanisms of resistance to penicillin and tetracycline produce larger zone diameters, which can be accurately, recognized using the interpretive criteria (Doern, 1995, John and James, 1999, Lalitha *et al.*, 2004).

Microbial resistance to conventional antibiotics has become a trending phenomenon. Over the years, scientists has always responded by discovery or production of a stronger antibiotics, combination of antibiotics or clinical rotation therapy to tackle the growing challenge and microorganisms at the receiving end has continually evolved ways of circumventing the adverse effect of antimicrobial agents beginning with their intrinsic nature. Resistance to antimicrobial agents typically occurs through five main mechanisms namely enzymatic inactivation of the drug, alteration of target sites, reduced cellular uptake, over production of neutralizing cellular components and extrusion by efflux (Salim *et al.,* 2005). Except for a few exemptions such as *S. aureus,* the study revealed that the Gram-positive isolates were more susceptible to the inhibitory effect of the antibiotics compared to their Gram-negative counterpart, which also corroborate the study by Salim *et al.* (2005) who reported lower MICs for *B. subtilis, S. mutans, S. aureus* compared to other gram-negative isolates. Some organisms are generally more susceptible than others depending largely on their structural framework (intrinsic resistance); some such as gram negative bacteria including *P. aeruginosa* are often more resistant than gram positive bacteria due to an outer phospholipid membrane with structural lipopolysaccharides (LPS) components that reduce the cell wall penetration ability of antimicrobial agents (Chessebrough 2006, Rahman *et al.,* 2011). Another example of intrinsic resistance that may have played up in the pattern of resistance is the chromosomally encoded constitutive β–lactamase enzyme that deactivates β–lactam antibiotics thereby conferring resistance on the organism. Resistance is said to be acquired or inducible when microorganisms become resistant to the activity of an antimicrobial agent to which it was previously susceptible. This may arise as a result of mutation or acquisition through horizontal gene transfer. Horizontal gene transfer involves acquisition of resistant genes through transformation, transduction, conjugartion, transposition or integron system. The result (see Table 3) showed that *S. aureus* isolate was resistant to the inhibitory effect of co-amoxiclav (30 µg), Erythromycin (15 µg), Gentamycin (120 µg), Trimethoprim (5 µg) and Doxycycline (30 µg). Similar pattern was observed for *P. aeruginosa* isolate which was resistant to the inhibitory effect of co-amoxiclav (30 µg), erythromycin (15 µg), sefoxitin (30 µg), trimethoprim (5 µg) and doxycycline (30 µg). Also *K. pneumoniae* isolate was resistant to the inhibitory effect of erythromycin (15 µg), trimethoprim (5 µg) and doxycycline (30 µg). This result is in line with results of studies done by Andrews (2001) who reported resistance of *S. aureus*, *K. pneumoniae* and *P. aeruginosa* clinical isolates to erythromycin, cefuroxime, vancomycin and nalidixic acid while the standard type strains remained susceptible. According to Salim *et al.* (2005), the phenomenon in which a bacterium is resistant to unrelated antibacterial agent is described as multi-drug resistance (MDR). One of the reason for the rapid emergence of resistant strains among clinical isolates may be attributed to the use of the versatile horizontal genetic transfer tool such as the bacterial plasmid. Generally, plasmids are extra chromosomal DNA capable of autonomous replication. R-plasmid are wide spread especially where antibiotic use is high such as hospitals and agricultural environment. R-plasmid was proved to be simultaneously resistant to four drugs which are Chloramphenicol, Sulphonamide, Streptomycin and Tetracyclines (Watanabe *et al.*, 2000, Salim *et al.*, 2005).

The results of minimum inhibitory concentration (MIC) is reflective of the pattern of IZDs even though it is more revealing of the fact that the isolates were relatively less susceptible than their standard type strains. In other words, all the isolates showed some degree of resistance compared to their standard type counterpart such that even the most susceptible isolate as shown in the result of this study (*B. subtilis*) recorded MICs more than four times higher than those recorded by its standard type counterparts (see Table 4). MIC is considered the 'gold standard' for determining the susceptibility of organisms to antimicrobials and are therefore used to judge the performance of all other methods of susceptibility testing (Andrews 2001, NCCLS 2003). MICs are used in diagnostic laboratories to confirm unusual resistance, to give a definitive answer when a borderline result is obtained by other method of testing. The MIC is the lowest concentration of antimicrobial agent that inhibits the growth of a particular organism and establishes the concentration of an antibiotic that is effective in preventing the growth of the pathogen. It also gives an indication of the dosage of that antibiotic that should be effective in controlling the infection in the patient (Andrews 2001, Firas *et al.,* 2008). By knowing the MIC and the theoretical levels of the antibiotics that may be achieved in body fluids, such as blood and urine, the physician can select an appropriate antibiotic, the dosage schedule, and the route of administration. Generally, a margin of safety of ten times the MIC is desirable to ensure successful treatment of the disease (NCCLS 2003, Lalitha *et al.*, 2004, Vinoth Kumar *et al.,* 2012). The lower the MIC value the more potent the antimicrobial agent, conversely, the higher the MIC value, the less potent the antimicrobial agent. Another revealing part of the results showed the interpreted resistant *S. aureus* and *P. aeruginosa* with MICs >10 times their standard type strains for erythromycin. Similar pattern was recorded for the fungal isolates as well.

4. CONCLUSIONS

All the clinical isolates were relatively less susceptible to the inhibitory effect of the antibiotics relative to the type strains. Aside S. aureus, all other gram-positive isolates were more susceptible to the inhibitory effect of the antibiotics. The two most resistant clinical isolates based on findings from this research are S. aureus and P. aeruginosa, followed by K. pneumoniae, E. coli and C. albican. Therefore, interventional efforts to limit the spread of antibiotic resistance must include periodic enlightenment of health personnel on:

- Innovative infection management strategies,
- Up to date antibiotics prescription guidelines,
- Reduction of non-professional handing of drugs,
- Enforcement of antibiotic usage policies,

- Reduced antibiotic use, clinical rotation of drugs,
- Exploitation of herbal options, and use of new and improved options such as vaccines, probiotics, interferons, immunomodulatory molecules, etc. for infection control.

ACKNOWLEDGEMENTS

The technical help of Dr. Onoriode and Dr. Obaze Osahon from Medical Microbiology Department, University of Benin Teaching Hospital, Benin City, Edo State, Nigeria; is appreciated and thankfully acknowledged.

REFERENCES

[1] Abraham, E.P. and Chain, E. (1940). An enzyme from bacteria able to destroy penicillin. *Review of Infectious Diseases* 10: 677– 678.

[2] Andrews, M.J. (2001). Determination of minimum inhibitory concentration. *Journal of Antimicrobial Chemotherapy* 48: 5–16.

[3] Bachman, A.M., LeBar, D. and Williams, E.L. (2017). First principle of clinical microbiology: Collection, handling and diagnostics. *Journal Biomedical Sciences* 3: 123–203.

[4] Doern, G.V. (1995). Susceptibility tests of fastidious bacteria. Manual of Clinical Microbiology, 6th edition, Murray P.R., Baron E.J., Pfaller, M.A., Tenover, F.C., Yolken, R., American Society for Microbiology, Washington D.C., 1342–1349.

[5] Firas, A., Bayati, L., Sulaiman, R. and Khudir, D.S. (2008). *In-vitro* antimicrobial activity of *Salvador apersica* extracts against some isolated oral pathogens in Iraq. *Turkish Journal of Biol*ogy 32: 57–62.

[6] John, D.T. and James, H.J. (1999). *Antimicrobial Susceptibility testing: General Considerations. Manual of Clinical Microbiology* 7th edition, Murray, P.R., Baron, E.J., Pfaller, M.A., Tenover, F.C., Yolken, R., American Society for Microbiology, Washington D.C., 1469–1473.

[7] Kreth, G.E., Jens, K.A., Merritt, J. and Qi, F. (2009). Bacterial and host interactions of oral Streptococci. *DNA and Cell Biology*, 28: 397–403.

[8] Lalitha, M.K., Manayani, D.J., Priya, L., Jesudason, M.V., Thomas, K. and Steinhoff, M.C. (2004). Manual on antimicrobial susceptibility testing. *Indian Journal of Medical Resources,* 106: 100–103.

[9] NCCLS-National Committee for Clinical Laboratory Standards (2003). *Performance Standards for Antimicrobial Disk Susceptibility and MIC Tests; Approved Standards.* 8th Edition. Information Supplement, Villanova. 23, M2–A8.

[10] Prescott, M.L., Harley, P.J. and Klein, D.A. (2005). *Microbiology* 6th edn., McGraw-Hill Inc., New York 990 pp.

[11] Rahman, M., Salehin, D., Jamal, O., Parvin, A. and Alam, P. (2011). Antibacterial activity of *Agemone mexicana* Leave extracts against waterborne microbes. *Research Journal of Medicinal plants,* 5: 621–626.

[12] Safa, A., Nair, G.B. and Kong, R.Y.C. (2010). Evolution of new variants of *Vibrio cholerae* O1. *Trends in microbiology,* 18: 46–54.

[13] Salim, A., Lan, R. and Reeves, P.R. (2005). Pathogenic Clones. *Emerging Infectious Diseases,* 11: 1758–1760.

[14] Usman, T.K., Al-Bagaz, J.I. and Einberg, D. (2005). Evaluation of some common agents used for controlling dental caries. *Microbios,* 139: 143–151.

[15] Vinothkumar, P., Sivaraj, A., Ahmed, K.S.Z., Sivamani, P., Devi, K. and Senthilkumar, B. (2012). Evaluation of antibacterial activities of *Andrographis paniculata* leaf extract against gram negative and gram positive species by *in-vitro* methods. *Journal of Pharmacological Research,* 3:1513–1515.

[16] Watanabe, T., Tani, M., Ishiba, Y.M. and Tokuyasu, K. (2000). Antibiotic susceptibility and plasmid profiling of selected clinical isolates. *Journal of Bioresources and Clinical Studies,* 3: 6710–6716.

[17] WHO (2010). Years lived with disability (YLDs) for 1160 sequelae of 289 diseases and injuries 1990–2010: A systematic analysis for the global burden of disease study. *Lancet,* 380: 2163–2196.

APPENDIX

Appendix Table 1: Antibiogram of the Selected Microbial Isolates

Organisms	*Zones of Inhibition (mean ± S.E. mm)*									
	AMC (30 µg)	*E (15 µg)*	*GN (120 µg)*	*CIP (5 µg)*	*FOX (30 µg)*	*W (5 µg)*	*DOX (30 µg)*	*KET (30 µg)*	*CLT (15 µg)*	*FLC (30 µg)*
B. subtilis	20 ± 1.0	19 ± 0.6	14 ± 0.3	27 ± 0.5	26 ± 0.3	15 ± 0.0	19 ± 1.2	–	–	–
S. pyogenes	18 ± 0.0	16 ± 1.5	15 ± 1.4	24 ± 0.0	23 ± 1.3	13 ± 0.5	16 ± 0.3	–	–	–
S. aureus	15 ± 1.2	9 ± 1.5	16 ± 1.6	23 ± 0.4	21 ± 0.0	11 ± 0.6	11 ± 0.5	–	–	–
E. coli	14 ± 0.3	6 ± 1.3	21 ± 1.0	19 ± 1.2	17 ± 0.3	11 ± 1.0	9 ± 1.5	–	–	–
P. vulgaris	17 ± 1.0	21 ± 0.6	16 ± 0.3	26 ± 1.5	22 ± 0.3	10 ± 1.0	15 ± 1.3	–	–	–
P. aeruginosa	5 ± 1.0	–	10 ± 0.0	17 ± 0.5	12 ± 1.5	–	6 ± 1.5	–	–	–
K. pneumonia	13 ± 0.3	6 ± 0.5	20 ± 1.7	18 ± 1.6	17 ± 0.6	6 ± 0.0	10 ± 0.6	–	–	–
C. neoformans	–	–	–	–	–	–	–	19 ± 1.5	22 ± 0.0	16 ± 0.6
C. albicans	–	–	–	–	–	–	–	17 ± 0.3	21 ± 0.6	12 ± 1.5

Keys: No inhibition (–), Amoxycillin-clavulanate (AMC), Erythromycin (E), Gentamycin (GN), Ciprofloxacin (CIP), Cefoxitin (FOX), Trimetoprim (W), Doxycycline (DOX), Ketokonazole (KET), Clotrimazole (CLT), Fluconazole (FLC).

Appendix Table 2: Interpretation of the Antimicrobial Sensitivity Pattern of the Isolates

Antimicrobial Agents	*Disc Conc. (in µg)*	*Interpretation of Results*								
		B. subtilis	*S. pyogenes*	*S. aureus*	*E. coli*	*P. vulgaris*	*P. aeruginosa*	*K. pneumoniae*	*C. neoformans*	*C. albicans*
Co-amoxiclav	30	S	S	R	I	I	R	I	–	–
Erythromycin	15	S	S	R	R	S	R	R	–	–
Gentamycin	120	I	I	R	I	I	I	I	–	–
Ciprofloxacin	5	S	S	S	S	S	I	S	–	–
Cefoxitin	30	S	S	I	I	S	R	I	–	–
Trimethoprim	5	S	I	R	I	I	R	R	–	–
Doxycycline	30	I	I	R	R	I	R	R	–	–
Ketokonazole	30	–	–	–	–	–	–	–	S	I
Clotrimazole	15	–	–	–	–	–	–	–	S	S
Fluconazole	30	–	–	–	–	–	–	–	I	R

Keys: S = Susceptible, **I** = Intermediate, **R** = Resistant, – = Not applicable

Appendix Table 3: Antibiogram of the Standard Type Microbial Strains

Organisms	*Zones of Inhibition (mean ± S.E., in mm)*									
	AMC (30 µg)	*E (15 µg)*	*GN (120 µg)*	*CIP (5 µg)*	*FOX (30 µg)*	*W (5 µg)*	*DOX (30 µg)*	*KET (30 µg)*	*CLT (15 µg)*	*FLC (30 µg)*
B. subtilis ATCC 21332	26 ± 1.0(S)	23 ± 0.6(S)	21 ± 0.3(S)	28 ± 0.5(S)	27 ± 0.3(S)	21 ± 0.0(S)	23 ± 1.2(S)	–	–	–
S. pyogenes ATCC12384	24 ± 0.0(S)	20 ± 1.5(I)	22 ± 1.4(S)	25 ± 0.0(S)	25 ± 1.3(S)	20 ± 0.5(I)	21 ± 0.3(S)	–	–	–
S. aureus ATCC 25923	22 ± 1.2(I)	19 ± 1.5(I)	24 ± 1.6(S)	29 ± 0.4(S)	25 ± 0.0(S)	18 ± 0.6(I)	22 ± 0.5(S)	–	–	–
E.coli ATCC 25922	25 ± 0.3(S)	20 ± 1.3(S)	25 ± 1.0(S)	27 ± 1.2(S)	21 ± 0.3(S)	19 ± 1.0(I)	17 ± 1.5(I)	–	–	–
P. vulgaris ATCC 29905	21 ± 1.0(S)	25 ± 0.6(S)	20 ± 0.3(S)	30 ± 1.5(S)	26 ± 0.3(S)	16 ± 1.0(S)	19 ± 1.3(S)	–	–	–
P. aeruginosa ATCC 29260	20 ± 1.0(S)	17 ± 1.3(I)	22 ± 0.0(S)	24 ± 0.5(S)	21 ± 1.5(S)	19 ± 0.4(I)	21 ± 1.5(S)	–	–	–
K. pneumonia ATCC 2342	21 ± 0.3(S)	18 ± 0.5(I)	24 ± 1.7(S)	23 ± 1.6(S)	25 ± 0.6(S)	20 ± 0.0(S)	22 ± 0.6(S)	–	–	–
C. neoformans ATCC 90112	–	–	–	–	–	–	–	25 ± 1.5(S)	28 ± 0.0(S)	22 ± 0.6(S)
C. albicans ATCC 10259	–	–	–	–	–	–	–	23 ± 0.3(S)	25 ± 0.6(S)	20 ± 1.5(S)

Keys: Susceptible **(S)**, Intermediate **(I)**, Resistant **(R)**, No inhibition (–), Amoxycillin-clavulanate (AMC), Erythromycin (E), Gentamycin (GN), Ciprofloxacin (CIP), Cefoxitin (FOX), Trimetoprim (W), Doxycycline (DOX), Ketokonazole (KET), Clotrimazole (CLT), Fluconazole (FLC).

Appendix Table 4: Minimum Inhibitory Concentrations (MICs) of the Selected Antibiotics against the Standard Type Strains and the Selected Isolates

Organisms	*MIC (in µg/ml)*									
	AMC	*E*	*GN*	*CIP*	*FOX*	*W*	*DOX*	*KET*	*CLT*	*FLC*
B. subtilis	0.2	0.5	1.2	0.08	0.1	1.4	1	–	–	–
B. subtilis ATCC 21332	0.07	0.1	0.08	0.03	0.06	0.3	0.8	–	–	–
S. pyogenes	0.4	0.7	1	0.09	0.4	1.8	1.1			
*S. pyogenes*ATCC12384	0.09	0.6	0.1	0.05	0.09	0.4	0.9			
S. aureus	1.1	>10	1.3	0.1	0.5	2.1	2			
S. aureus ATCC 25923	0.5	0.8	0.3	0.08	0.1	1.1	1			
E. coli	1.2	1.8	0.6	0.6	0.8	3	2.9	–	–	–
E. coli ATCC 25922	0.8	1	0.1	0.1	0.2	1.2	1.3			
P. vulgaris	0.5	0.3	1.2	0.2	0.3	2	1.2			
P. vulgaris ATCC 29905	0.1	0.06	0.1	0.05	0.1	0.7	0.9			
P. aeruginosa	2.3	>10	1.5	1.3	1	5	3.5	–	–	–
P. aeruginosa ATCC 29260	0.9	1	1	0.8	0.6	1	0.8			
K. pneumoniae	1.2	1.8	0.4	0.5	0.7	3	3	–	–	–
K. pneumonia ATCC 2342	0.8	1	0.1	0.1	0.2	1.2	1.3	–	–	–
C. neoformans	–	–	–	–	–	–	–	3.4	1.2	6
C. neoformans ATCC 90112	–	–	–	–	–	–	–	0.9	0.6	1.5
C. albicans	–	–	–	–	–	–	–	5.2	2	>10
C. albicans ATCC 10259	–	–	–	–	–	–	–	1.1	0.8	3

Keys: > = Greater than – = Not applicable, Amoxycillin-clavulanate (AMC), Erythromycin (E), Gentamycin (GN), Ciprofloxacin (CIP), Cefoxitin (FOX), Trimetoprim (W), Doxycycline (DOX), Ketokonazole (KET), Clotrimazole (CLT), Fluconazole (FLC).

Identification of *Acinetobacter* spp. Isolated at the End of the Civil War in Sri Lanka and their Antimicrobial Susceptibility Status

Darshana Wickramasinghe[1*], Sujatha Pathirage[1], Shirani Chandrasiri[2], Geethika Patabendige[3] and Renuka Rajapakshe[4]

[1]National Institute of Health Sciences, Kalutara, Sri Lanka
[2]Colombo South Teaching Hospital, Kalubowila, Sri Lanka
[3]Cancer Institute, Maharagama, Sri Lanka
[4]Medical Research Institute, Colombo, Sri Lanka
E-mail: *dwickw2@yahoo.co.uk

ABSTRACT: The objective was to identify the Acinetobacter spp. isolated in Teaching Hospital, Anuradhapura, Sri Lanka by phenotypic tests and to study antimicrobial susceptibility status of those isolates. This hospital based descriptive study consisted of Acinetobacter spp. positive, clinically significant culture samples received at a microbiology laboratory, in a tertiary care hospital in Sri Lanka from 1st of August to 31st October 2009. All the Acinetobacter isolates were confirmed and subjected to simplified phenotypic tests. Assimilation tests were performed and antimicrobial susceptibility was determined by the Clinical and Laboratory Standards Institute (CLSI) guidelines, 2008. From the clinically significant Acinetobacter isolates, 7% were from the wards, 26.7% were from pediatric intensive care unit, 10% were from premature baby unit, and 46.7% were from intensive care unit. Acinetobacter were isolated from respiratory tract secretions (63.3%), blood (26.7%), urine (3.3%) and wound exudates (6.7%). With the simplified phenotypic identification testing, the 30 isolates from the study population belonged to only 5 phenotypes (phenotype numbers 1, 2, 5, 12, 13). Phenotype 2 (A. baumannii) is the predominant species (56.7%) followed by Phenotype 1 (A. calcoaceticus) 23.3%, Phenotype 5 (A. junii) 10%, Phenotype 13 (Acinetobacter genomosp. 13) 6.7% and Phenotype 12 (A. radioresistens) 3.3%, respectively. Resistance to Cephalosporins (> 73%) was very high. Resistance to Imipenem was

70%. When considering about aminoglycosides, amikacin showed the best sensitivity (47%). Levofloxacin had a better sensitivity (40%) than Ciprofloxacin (27%). Co-trimoxazole showed 77% resistance. 53% of the isolates were cefoperazone - sulbactam resistant but only 20% gave the complete resistance. Tigecycline (90%) and colistin (100%) are the only drugs which gave good sensitivities. A simple identification scheme is useful for laboratories with limited resources for typing Acinetobacter isolates. Majority of the isolates were multi-drug resistant (70%).

Keywords: Civil War, Sri Lanka, Antimicrobial Susceptibility.

1. INTRODUCTION

Acinetobacter spp. was first described in 1911 as *Micrococcus calcoaceticus.*[1] The genus *Acinetobacter* is classified under the family Moraxellaceae and comprises strictly aerobic, gram-negative, non motile, non-lactose-fermenting, oxidase negative, catalase positive, coccobacillus. *Acinetobacter* spp. as a human pathogen has been reported sporadically. Currently, at least 32 different DNA groups have been reported. They are ubiquitous, free-living and fairly stable in the environment. They are commonly isolated from water, soil, foods, arthropods etc.[2] In humans, *Acinetobacter* can colonize skin, wounds, and the respiratory and gastrointestinal tracts. These are natural reservoirs and account for up to 35–45% of bacteria isolated from human skin.[3]

Acinetobacter speciation by a simplified phenotypic identification has not been done in Sri Lanka. In the present study, an attempt was made to type the *Acinetobacter* isolates obtained from clinical samples by a simplified phenotypic identification scheme with a numerical approach and also to determine their antimicrobial susceptibility.

2. METHODOLOGY

This study was a hospital based descriptive study. The study samples consisted of *Acinetobacter* spp. positive, clinically significant culture samples received at the Microbiology Laboratory, Teaching Hospital Anuradhapura, Sri Lanka from 1st of August to 31st October 2009. All *Acinetobacter* isolates were confirmed, subjected to simplified phenotypic tests as described by Gerner-Smidt *et al.* (1991)[5] and Bouvet *et al.*

(1989).[6] The assimilation tests were performed as described by Stainer *et al.* (1966).[7]

Determination of antibiotic susceptibility (ABST) was done using the standard disk diffusion method recommended by the Clinical and Laboratory Standards Institute and zone diameter interpretation was done according to the zone diameter interpretive standards of the CLSI guidelines, 2008. Reading of zone diameter of Ceperperozone-Sulbactum was done according to the Sulperazone Manual for the Microbiologists, Pfizer Pharmaceutical Division, Pfizer Infection Control, USA (1988). The research protocol was approved by the Colombo South Teaching Hospital, Ethical Review Committee.

3. RESULTS

From the isolates during the period of 3 months, 3% were found as clinically significant *Acinetobacter* isolates. 16.7% of the *Acinetobacter* isolates were from the wards, 26.7% were from paediatric intensive care unit, 10% were from premature baby unit, and 46.7% were from intensive care unit (Figure 1). *Acinetobacter* were isolated from respiratory tract secretions (63.3%), blood (26.7%), urine (3.3%), and wound exudates (6.7%).

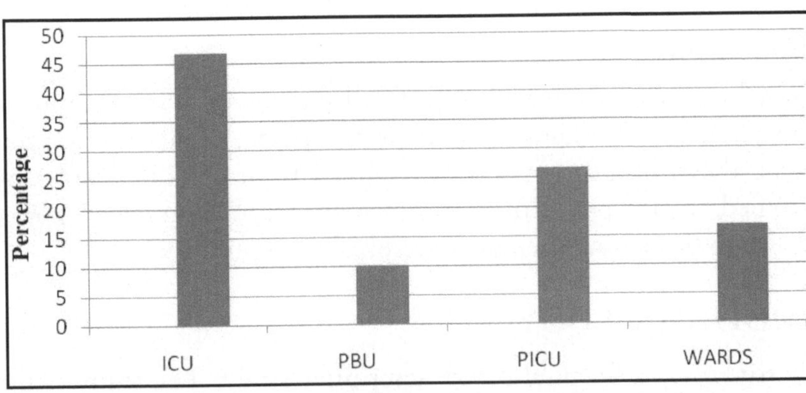

Figure 1: Distribution of the Location of the Patients Harboring *Acinetobacter* Isolates

With the simplified phenotypic identification testing, 30 isolates from the study population belonged to only 5 phenotypes (phenotype numbers 1, 2, 5, 12, 13). Phenotype 2 (*A. baumannii*) is the predominant species (56.7%) followed by Phenotype 1 (*A. calcoaceticus*)

23.3%, Phenotype 5 (*A. junii*) 10%, Phenotype 13 (*Acinetobacter* genomosp.13) 6.7% and Phenotype 12 (*A. radioresistens*) 3.3%, respectively (Table 1).

Table 1: Species Differentiation of the Isolates and their Distribution in Clinical Specimens

Acinetobacter spp.	*Geno Species*	*Sample Numbers*				*Number*
		Blood	*Respiratory*	*Urine*	*Pus*	
Acinetobacter calcoaceticus	1	2	3, 5, 18, 19, 20, 30			7 (23.33%)
Acinetobacter baumannii	2	1, 17, 25, 27	6, 7, 8, 10, 11, 13, 14, 15, 16, 23, 24, 28	9		17 (56.66%)
Acinetobacter Junii	5	21, 26			12	3 (10%)
Acinetobacter radioresitens	12				29	1 (3.33%)
Acinetobacter genomosp.13	13	4	22			2 (6.66%)

Majority of the isolates (70%) were multi-drug resistant (MDR), showing resistance to two or more antimicrobial agents. From all *Acinetobacter baumannii* isolates, 94.11% (16) were MDR. Tigecycline shows 85.7% and Colistin 100% sensitivity against MDR *Acinetobacter* isolates.

Resistance to cephalosporins (> 73%) was very high. Resistance to Imipenem was 70%. When considering about aminoglycosides, amikacin showed the best sensitivity (47%). Levofloxacin had a better sensitivity (40%) than ciprofloxacin (27%). Cotrimaxazole showed 77% resistance. 53% of the isolates were Cefoperazone Sulbactam resistant but only 20% gave the complete resistance. Tigecycline (90%) and colistin (100%) are the only drugs which gave good sensitivities (Table 2).

Table 2: Antimicrobial Susceptibilities of *Acinetobacter* Isolates

Antibiotic	*Percent Resistant/Intermediate*
Cefotaxime	90
Coamoxiclav	73.3
Ceftazidime	73.3
Imipenem	70
Amikacin	53.4
Gentamicin	73.4
Ciprofloxacin	73.4
Ceperferazone - sulbactam	53.4
Tigecycline	10
Piperacillin - tazobactam	70
Polymyxin E (Colistine)	0
Levofloxacine	60
Cefepime	73.3
Co-trimozaxole	76.6

4. DISCUSSION

Acinetobacter species has emerged as an important nosocomial pathogen that is often MDR and associated with life-threatening infections.[4]

The Phenotypic identification of *Acinetobacter* to the species or DNA group level by Bouvet & Grimont (1987)[8] seems to be appropriate but a large panel of phenotypic tests is tedious and time consuming to perform.[4] Simple identification schemes such as those of Gerner-Smidt & Ursing (1991);[5] Tjernberg (1990)[9] alone with antimicrobial susceptibility testing may be useful for laboratories with limited resources and can be adapted for typing isolates, thus avoiding the need for expensive molecular methodology.

Identification of *Acinetobacter* species based upon growth at 44°C, 41°C and 37°C in BHI, penicillin G (10 µg) and chloramphenicol (30 µg) sensitivity and acid production from glucose could be very useful. In addition, only 5 assimilation tests were used to differentiate isolates upto species level.[5] Apart from that, isolates were tested for haemolysis on sheep and human blood (5%) agar and gelatin liquefaction. According to the literature, if there are haemolytic isolates we have to do additional carbon source assimilation tests to differentiate those isolates upto species level.[9] Simple identification scheme with few phenotypic tests would be a better alternative method for the resource poor laboratories in countries like Sri Lanka.

Phenotypic characteristics of *Acb*-complex (DNA groups 1, 2, 3, 13) were helpful in differentiating the species within the complex. Most studies they have discussed the problems in separating DNA groups 1, 2, 3 and 13 by phenotypic tests. The responses for most of the assimilation tests were very similar for DNA groups 1, 2 and 13. This similarity is in accordance with the observations that these four groups are genotypically more closely related to each other than to other DNA groups (Tjernberg & Ursing 1989).[10]

However, there were no unequivocal tests separating DNA group 2 from DNA group 13, which was not recognized by Bouvet and Grimont (1987).[8] But in this study, isolates susceptible to both penicillin (10 μg) chloramphenicol (30 μg) were used to separate those two groups as in case of Tjernberg (1990).[9]

According to the literature, hemolytic strains were found in DNA groups 4, 5, 6 and 14.[8] For all of these groups except group 5, hemolysis of human and sheep blood correlated with a positive gelatinase test. Even though the isolates were tested for haemolysis on sheep and human blood (5%) agar and gelatin liquefaction, there were no positives in this study. So, additional carbon source assimilation tests were not performed.

The tests for identification of Acinetobacters selected by Bouvet and Grimont (1987)[8] seem appropriate for the DNA groups, which were isolated in this study. As further continuation of this study, genotypic identification was done for 24 isolates out of 30 and they tally with the phenotypic identification results.

The literature search demonstrates that genomic species Gs-2 (*A. baumannii*), together with Gs-1 (*A. calcoaceticus*), Gs-3 and Gs-13TU, is predominantly involved in infection and is often collectively known as the *Acinetobacter calcoaceticus-A. baumannii* (*Acb*) – complex (Gerner-Smidt 1991).[5] This holds true in our study too and 86.66% of the isolates belonged to *Acinetobacter calcoaceticus-A. baumannii* (*Acb*) – complex.

Majority of the species isolated in this study (Teaching Hospital, Anuradhapura) were *A. baumannii* (56.66%). According to the past studies, this is the main species responsible for nosocomial infection in other parts of the world too[11]. In this study, majority of the isolates were multi-drug resistant (70%), showing resistance to two or more antimicrobial agents. From all *Acinetobacter baumannii* isolates, 94.11% (16) were multi-drug resistant.

MDR *Acinetobacter* isolates show 85.7% sensitivity to Tigecycline and 100% to Colistin but 100% resistant to Cephalosporins and Carbapenems and > 80% resistance to all the other antibiotics. Tigecycline resistance was noted only in *Acinetobacter baumannii.*

A high efficiency of ampicillin/sulbactam is worth considering. Sulbactam is known to be much more efficient against Acinetobacters than Clavulanic acid which is closely related in its affinity to PBP.[12] It has to be pointed out that the opinions about the efficacy of beta-lactamase inhibitor tazobactam rather differ.[13] As ampicillin/sulbactam discs were not available during the study period, sensitivities were checked only for ceperferozone sulbactam. In this study, ceperferozone sulbactam shows fairly good sensitivity comparing to other group of drugs (sensitivity 46.6%, intermediate sensitive 33.3%, resistant 20%).

The resistance of Acinetobacters to beta-lactam antibiotics can be due to basic mechanisms, namely enzyme destruction through beta-lactamases, alterations of target sites and reduced permeability of bacterial walls.[14]

The development of resistance to antimicrobials in *Acinetobacter* appears to be unstoppable, when we compare the study reports publicshed over time.[4] Polymyxin B and colistin have demonstrated a reasonable success in controlling MDR *Acinetobacter* in the past. But, used only for selected cases of serious nature due to their higher toxicities. In our study, Colistin is the only drug that gave 100% sensitivity. Apart from colistin, tigecycline is the only alternative with 85.7% sensitivity against MDR *Acinetobacter* isolates.

Many institutions around the world are faced with the challenging issue of pandrug resistance. Concerns have been raised about the use of Tigecycline for *A. baumannii* infection, particularly for bacteremia, leaving colistin as the only therapeutic option for some.

5. CONCLUSION

With the simplified phenotypic identification testing, 30 isolates in the study population belonged to only 5 phenotypes (phenotype numbers 1, 2, 5, 12, 13). Phenotype 2 (*A. baumannii*) is the predominant species (56.66%) followed by Phenotype 1 (*A. calcoaceticus*) 23.33%. Simple identification scheme is useful for laboratories with limited resources for typing *Acinetobacter* isolates. Majority of the isolates were multi-drug resistant (70%).

ACKNOWLEDGEMENTS

1. Department of Microbiology, National Institute of Health Sciences, Kalutara.
2. R.N. Renuka Rajapakshe, Medical Research Institute, Colombo.

REFERENCES

[1] Beijernck, M.W., Uber Pigmentbildung bei Essigbakterien (1911) Cent Bakteriol Parasitenk 29:169–76.

[2] Fournier, P.E. and Richet, H. (2006). The epidemiology and control of *Acinetobacter baumanii* in healthcare facilities. *Clinical Infectious Diseases,* 42:692–9.

[3] Patil, J.R. and Chopade, B.A. (2001). Distribution and *in vitro* antimicrobial susceptibility of *Acinetobacter* species on the skin of healthy humans. *National Medical Journal of India,* 14(4): 204–208.

[4] Towner, K.J. (1997). Clinical importance and antibiotic resistance of *Acinetobacter* spp. *Journal of Medical Microbiology,* 6:186–97.

[5] Gerner-Smidt, P., Tjernberg, I. and Ursing. J. (1991) Reliability of phenotypic tests for identification of *Acinetobacter* species. *Journal of Clinical Microbiology,* 29:277–282.

[6] Bouvet, P.J. and Grimont, P.A. (1989). Taxonomy of the genus *Acinetobacter* with the recognition of *Acinetobacter baumannii* sp. nov., *Acinetobacter haemolyticus* sp. nov., *Acinetobacter johnsonii* sp. nov., and *Acinetobacter junii* sp. nov., and amended description of *Acinetobacter calcoaceticus* and *Acinetobacter lwoffii. International Journal of Systematic and Evolutionary Microbiology,* 36:228–240.

[7] Stainer, R.Y., Palleroni, N.I. and Doudoroff, M. (1966). The aerobic pseudomondads: a taxonomic study. Journal of General Microbiology 43:159–271.

[8] Bouvet, P.J. and Grimont, P.A. (1987). Identification and biotyping of clinical isolates of *Acinetobacter. Ann. Inst. Pasteur Microbiol,* 138:569–578.

[9] Tjernberg, I. (1990). Antimicrobial susceptibility of *Acenetobacter* spps identified by DNA-DNA hybridization. Acta Pathologica, *Microbiologica et Immunologica Scandinavica,* 98: 320–326.

[10] Tjernberg, I. and Ursing, J. (1989). Clinical strains of *Acinetobacter* classified by DNA-DNA hybridization. Acta Pathologica, *Microbiologica et Immunologica Scandinavica,* 97:595–605.

[11] Houang, E.T., Chu, Y.W., Leung, C.M., Chu, K.Y., Berlau, J., Ng, K.C. and Cheng, A.F. (2001). Epidemiology and infection control implications of *Acinetobacter* spp. in Hong Kong. *Journal of Clinical Microbiology,* 39:228–234.

[12] Vila, J., Marcos, A., Marco, F., Abdalla, S., Vergara, Y., Reig, R., Gomez-Lus, R. and Jimenez de Anta, T. (1993). In vitro antimicrobial production of -lactamases, aminoglycoside-modifying enzymes, and chloramphenicol acetyltransferase by and susceptibility of clinical isolates of *Acinetobacter baumannii. Antimicrobial Agents Chemotherapy,* 37:138–141.

[13] Joly-Guillou, M.L. and Decre, D. (1995). Bactericidal *in vitro* activity of ß-lactams and ß-lactamase inhibitors, alone or associated, against clinical strains of *Acinetobacter baumannii*: effect of combination with aminoglycosides. *Journal of Antimicrobial Chemotherapy,* 36: 619–629.

[14] Dorsey, C.W., Beglin, M.S. and Actis, L.A. (2003). Detection and analysis of iron uptake components expressed by *Acinetobacter baumannii* clinical isolates. *Journal of Clinical Microbiology,* 41:4188–4193.

Food as a Vehicle for Transmission of Antimicrobial Resistance

Harsh Panwar*, Rohini Devidas Gulhane and Niharika Singh

Department of Dairy Microbiology, College of Dairy Science and Technology, Guru Angad Dev Veterinary and Animal Sciences University (GADVASU), Ludhiana (Punjab), India

E-mail: *drhpanwar@gmail.com

ABSTRACT: Food serves as the principle source of nutrients and energy for humans and animals. Besides, energy and nutrition, magnitude of microflora present in foods is taken as an index of food microbiological quality and as a parameter for adjudging the overall food acceptance. However, the antibiotic resistance profile of the resident strains is not much emphasized. Food, primarily of animal origin can serve as a vehicle for transmission of resistant strains and antimicrobial resistance traits. The resident flora of milk and milk products generally recognized as safe (GRAS) can also transfer resistance traits to strains of clinical significance. Microbes are an integral part of human system and out-number the human cells. Attempts are made to either restrict the number of microbial cells in food, or otherwise to supplement or fortify food with selective microbial strains. Several processing conditions are applied to reduce their number to an acceptable limit laid by regulatory authorities. However, microbes have enormous potential to adapt to a wide variety of stress conditions. Emerging stress resistance offers a challenge to control of microbes in food. Food matrix further protects bacteria from gastrointestinal stress factors and helps in its safe delivery to host intestine, resulting in infection or toxico-infection. Current AMR mitigation strategies viz. application of bacteriophages, vaccines, bacteriocins etc. as alternate to antibiotics, and genetic strategies have limited application or acceptability in food. Application of antimicrobial peptides primarily of milk origin and their dynamic variants, referred to as DAMPs have shown promise in addressing AMR in food. Further, globalization and establishment of a global market has further augmented the free mobility of antimicrobial resistant strains and probability of food crises and incidents. Thus, there is an urgent need for a global response plan to design mitigation strategies and practices for control of multi-drug resistance strains and their transmission thereof by the implementation of an effective system of food risk intelligence on a global scale.

Keywords: Food, Milk, Antimicrobial Resistance, Probiotic, Lactic acid Bacteria, Antimicrobial Peptides.

1. INTRODUCTION

Microorganisms are closely associated with human beings and influence different processes significantly including human health, ecosystem sustainability, and several industrial processes. Humans and other mammals colonize a complex spectrum of microorganisms. Adult human body is home to a greater number of prokaryotic cells than body's own cells, with estimated 90% of microbial cells and only 10% of eukaryotic cells. However, the decade old assumption has been busted by a recent report published in Nature, which states that the ratio between resident microbes and human cells is more likely to be one-to-one. This microbiota makes up a huge biomass of > 100,000 billion bacteria encompassing over > 400 different species. Most of these bacterial cells lie within human intestine, where their impact is likely to be the most through influencing metabolic activity. Manipulation of these gut microbial communities is explored for management of several diseases and disorders; primarily metabolic disorders *i.e.* obesity, diabetes, and cardiovascular diseases. Microorganisms also play important role in several food chains upon which all life depends. Foods, whether of plant or animal origin are rarely if, ever sterile. The type of microorganisms associated with foods depends upon the mode of entry, processing conditions, type of organisms gaining access, nutritional status of food and interaction of microorganisms with food components over the time. The presence of microorganisms in raw material and processed food shapes the properties, storage conditions, shelf life and overall acceptability of the food products. In some cases, primarily in context to fermented foods, the selected microflora has no discernible effect on food quality and acceptability, and the food is consumed freely. The presence of microbial strains in fermented foods contributes to the organoleptic characteristics *i.e.* aroma, savour, texture, shelf-life, and bio-therapeutic potential of food. These strains also add on to natural antimicrobials like organic acids, hydrogen peroxide, bacteriocins, as well as uncharacterized proteinaceous and phenolic substances. However, in some other cases, microorganisms reflect their presence as spoilage and disease outbreaks.

In the current scenario, the major emphasis is on the evolving anti-microbial resistance profile of microbial strains, primarily of pathogenic nature. The resistant strains, like other microorganisms are mobile and travel across different ecological niches and spread resistance determinants to other strains. Hospital environment and animal farms are generally linked to emergence and spread of resistant strains, however, recently food has also been witnessed as a mode of transfer of resistance strains. Food, primarily of animal origin can serve as a vehicle for transmission of resistant strains and antimicrobial resistance traits. Besides the microbial inhabitants of food, it can lead to the transmission of resistance of genetic determinants to other commensal and pathogenic strains. Globalization has further augmented the free mobility of antimicrobial resistant strains and probability of food crises and incidents. Thus, there is an urgent need for a global response plan to design mitigation strategies and practices for control of multi-drug resistance strains and their transmission thereof by the implementation of an effective system of food risk intelligence on a global scale.

2. FOOD AND MICROBES

Hippocrates, nearly 2500 years ago stated "Let food be thy medicine and medicine be thy food" with a view that good nutrition plays a critical part in promoting health and disease prevention. Micro-organisms have multi faceted role in food processing, fermentation and towards food security for mankind. Many beneficial bacterial, yeast and mould strains are widely applied to convert raw food substrates into fermented products *i.e.* curd, cheese, kefir, kumiss, rabdi *etc.* Lactobacilli are among the most preferred and explored strains. The genus *Lactobacillus* is attracting much interest due to its ever-increasing beneficial effects and functional properties. They are also being employed in animal feed as replacer for antibiotic growth promoters. These beneficial strains are being referred to as probiotics *i.e.* the ***living microorganisms which when administered in adequate amounts confer some health benefits to the host*** (FAO/WHO, 2002). These strains have acquired '***generally recognized as safe (GRAS)***' status and are administered through fermented foods and pharma products. Multiple reports support safety and efficacy of specific strains of probiotics under *in-vitro*, *in-vivo* and human clinical trials.

2.1 Pathogenic Microbes in Food

Besides beneficial strains in food, foods can also harbour several commensal and pathogenic strains. Commensal strains make their way from raw material microflora, processing conditions, packaging material and during storage and generally, do not pose any risk to human health. Pathogenic strains can enter into food chain directly from animal sources and particularly through untreated fruits, vegetables, raw meat products, raw milk products as well as water sources. Contamination of food by pathogenic strains poses a threat to consumer resulting into infection or intoxication. Food associated pathogens can be zoonotic, environmental or of human origin. The epidemiology of microbe mediated food-borne illnesses has changed due to increasing susceptibility of human population to diseases and also due to emergence of new evolved microbial pathogens. Food-borne pathogens of high concern are *Campylobacter jejuni*, *Escherichia coli* O157:H7, *Listeria monocytogenes*, *Salmonella* spp., *Listeria* spp., *Cronobacter sakazakii,* etc. Milk and milk products can easily come across these pathogens and can be important sources of food-borne pathogens for consumers.

3. FOOD PROCESSING PARAMETERS AND MICROBIAL STANDARDS

The emergence of food borne pathogens has led to increased awareness and several regulations regarding the maximum permitted limits have been implemented. ***FSSAI (Food Safety and Standards Authority of India)*** has laid microbiological standards for milk, milk products and other food products that defines the quality of food items. Food processing parameters are implemented to limit the microbial counts below the given standards. The maximum permissible limits for different pathogens in milk and milk products and meat products have been summarized in Table 1.

Table 1: Maximum Permissible Limits for Pathogens in Milk and Milk Products and Meat Products

Product	*Aerobic Plate Count*	*Coliform*	*S. aureus*	*Yeast and Mold*	*E. coli*	*Salmonella/ Shigella*	*L. monocytogenes*
Pasteurized/ Fermented milk	50×10^3	Absent/0.1 ml	–	–	Absent/0.1 ml	Absent/25 ml	Absent/25 ml
Pasteurized cream	75×10^3	10/g	–	–	Absent/0.1 g	Absent/25 ml	Absent/25 ml
Sweetened condensed milk	5×10^2	10/g	Absent/0.1 g	10/g	Absent/0.1 g	Absent/25 g	Absent/g
Milk powder	30×10^3	50/g	100/g	5/0.1 g	Absent/0.1g	Absent/25 g	Absent/25 g
Infant milk food	5×10^2	Absent/0.1 g	Absent/0.1 g	Absent/0.1 g	Absent/0.1 g	Absent/25 g	Absent/25 g
Ice-cream	1×10^3	100/g	100/g	–	Absent/0.1 g	Absent/25 g	Absent/g
Fresh meat	5×10^6	–	100/g	50×10^3	500/g	Absent/25 g	Absent/25 g
Chilled meat	5×10^6	–	100/g	10×10^3	100/g	Absent/25 g	Absent/25 g
Frozen meat	1×10^5	–	100/g	1×10^3	10/g	Absent/25 g	Absent/25 g

4. FOOD AND ANTIMICROBIAL RESISTANCE

Antimicrobial resistant (AMR) microbes present on food constitute a direct danger to global public health. However, antimicrobial resistance genes in pathogenic or commensal strains form an indirect public health challenge. Food can be infected with AMR bacteria and/or antimicrobial resistance genes in several manners. A foremost way is the occurrence of AMR bacteria on the food selected by the adoption of antibiotics during agricultural production. A second route is the probable presence of resistance genes in bacteria that are intentionally supplemented during the processing of food (starter cultures, probiotics etc.). A final pathway is through cross-contamination with AMR bacteria during the different stages of food processing. A strong risk for transmission of antimicrobial resistance to humans is mainly due to consumption of raw or uncooked food products without having undergone prior processing or preservation. As a result, transfer of antimicrobial resistance genes among bacteria after ingestion by humans may occur. The processes involved in food production could kill bacteria and decrease the risk of transmission of antimicrobial resistance. The human gastrointestinal (GI) tract is a complex eco-system with a broad species diversity, of at least 400 different bacterial species (Borriello 1986). Despite many attempts to regulate the antibiotic resistance development, the level of resistant bacteria is on the upgrade and the speculation that our GI tract acts as a reservoir of antibiotic resistance genes is universally accepted (Grasselli *et al.*, 2008). The microflora or pH of the human colon, usually considered beneficial, could be playing a more sinister factor in the development and spread of AMR bacteria. AMR has been described in various different species of bacteria. Methicillin-resistant ***Staphylococcus aureus (MRSA)*** is one of the best-known illustrations. The other most common examples are ***Vancomycin-resistant Enterococcus (VRE), Multi-drug-resistant Mycobacterium tuberculosis (MDR-TB), Carbapenem-resistant Enterobacteriaceae (CRE)*** gut bacteria. The transmission of MRSA from food animals to humans is well recognized and transfer of MRSA through the food chain has also been documented (Hanselman *et al.,* 2006, Lewis *et al.,* 2008).

4.1 Hong Kong Strategy and Action Plan on Antimicrobial Resistance (2017–2022)

With regard to the presence of AMR bacteria in foods, it is significant to prevent humans from getting infested with these bacteria. In this regard, the Government of Hong Kong has developed and launched 'Hong Kong Strategy and Action Plan on Antimicrobial Resistance (2017–2022)'. According to that Strategic action plan 5 preventive keys were identified for food contaminations. Those are:

- Choose safe raw materials
- Keep hands and utensils clean
- Separate raw and cooked food
- Cook thoroughly and
- Keep food at safe temperature.

The document also emphasised that everybody should pursue to prevent food borne diseases, including food borne infections due to AMR bacteria.

5. MICROBES AND ANTIMICROBIAL RESISTANCE

Antimicrobial resistance is the competence of a microbial strain to resist the growth inhibition or killing action of an antibiotic beyond the normal susceptibility of the specific bacterial strain. This resistance towards antibiotics may be due to several microbial competencies *i.e.* enzymatic degradation of antibiotics, modification of antibiotic target, changing the cell membrane permeability and continuous efflux of antibiotics. Possibility of gene transfer between bacterial strains is an increasing public health concern. Food can not only lead to transmission of commensal microorganisms between different ecological niches but also food borne pathogens. Microorganisms can spread from food to humans during handling of raw materials as well as through cross- and re-contamination between different food products. Upon consumption, these pathogenic bacteria pose an instant threat to public health. The effects of multi-drug resistant *Salmonella* and *Campylobacter* spp. for consumers have been studied frequently. The reports revealed that the emerging resistance among these food borne pathogens leads to increased number of hospitalizations and rise in invasive infections and mortality. Presence of bacterial DNA carrying genetic resistance determinants in food products pose an indirect danger to public health, as it results into rise in gene pool from which commensal or pathogenic bacteria pick up these resistance determinants and possibly transfer them to other pathogenic bacteria. Several *in-vitro* studies have demonstrated the transfer of resistance genes from beneficial and commensal strains to pathogenic strains. Resistance to antibiotics in commensal strains constitutes an indirect risk to public health as antimicrobial resistance genes may be transferred to pathogens. For example, pathogenic strains consumed through food may contain resistance genes located on mobile genetic elements. It is thus possible that resistance gets transferred to pathogens in the human GIT. During the production process of certain food products, microorganisms are intentionally added *e.g.* starter cultures, probiotic, bio-preserving microorganisms and bacteriophages. Transferable resistance genes are rare in starter cultures, while resistance is most common in *Enterococcus* spp. In most cases, *Enterococcus* is resistant to vancomycin, although resistances to tetracycline, erythromycin and chloramphenicol has also been observed.

6. PROCESSING PARAMETERS AND RESISTANCE

Processing parameters employed for food preparation and for minimizing the microbial number below the permissible limits results in resistance and adaptation of microbial strains towards the processing hurdles. Food processing parameters, preservation/storage stress, gastrointestinal stress, and body defence mechanisms significantly impacts the microbial strains. Several studies relate processing parameters/stress to emerging antimicrobial resistance. In one such earlier study, Lou and Yousef (1997) studied the effect of sub-lethal environmental stress on *L. monocytogenes.* Cells in exponential phase were allowed to grow for 1 h at a sub-lethal dose of ethanol (5%, *v/v*), acid (HCl, pH 4.5 to 5.0), hydrogen peroxide (500 ppm), heat (45°C) and NaCl (7%, *w/v*). Adaptation to pH and ethanol significantly ($P < 0.05$) increased *L. Monocytogenes* resistance to lethal doses of acid, ethanol, and H_2O_2. Upon adaptation to H_2O_2 (500 ppm), resistance of the pathogen increased to 1% H_2O_2. Heat shock significantly increased the resistance to ethanol and NaCl. Therefore, 'stress hardening' should be considered when current food processing technologies are modified or new ones are developed. Another study evaluated the acid and heat resistance of lactic acid adapted *Salmonella enteritidis* in gastric fluid (pH 2.0) and during thermal treatment (54–60°C). Lactic acid adapted cells were shown to have higher acid and heat resistance and less fluid membrane than control cells. The increased acid and heat resistance of lactic acid adapted *S. enteritidis* was attributed to the decreased membrane fluidity (Yang *et al.*, 2014). In an earlier study by Pleitner and co-workers (2012), sub-lethal heat shock to *E. coli* improved its survival at 60°C. Presence of NaCl was also reported to augment heat resistance. Cells were reported to accumulate carbohydrates in presence of 6% NaCl. The developed heat resistance in *E. coli* was attributed to ribosome stability due to accumulation of cytoplasmic solutes. Recent findings documented that when bacteria are exposed to sub-lethal heat stress and UV, they develop a pharesistant phenotype, however, bacteria subjected to sub-lethal chlorine does not develop this resistance. Findings suggest that bacteria surviving sub-lethal disinfection conditions might have higher survival chances directly after treatment compared to un-challenged cells (Ameh *et al.*, 2018).

7. MITIGATION STRATEGIES

Different mitigation strategies are employed for management of multi-drug resistant strains. However, for the strategies to be employed in food, the strategy should be of GRAS nature and must not impact textural and organoleptic properties of food products. The different novel strategies are as follows.

7.1 Probiotics, Prebiotics and Synbiotics

The ban of dietary antimicrobial agents as growth promoters attracted attention towards use of probiotics, prebiotics and synbiotics for production of safe food with ensured gut health. Probiotics and *Direct-Fed Microbials (DFMs)* have been used as animal feed that serves as an alternative to current-day antibiotics. Several reports have proposed application of probiotic strains as growth promoters as a substitution of antibiotics (Munoz-Atienza *et al.*, 2013, Tellez *et al.*, 2015). Probiotics helps in maintaining the gut microbial balance, besides acting as an immune-stimulant. Probiotics acts by stimulating healthy intestinal microbial ecosystem, promoting digestion, nutrient absorption and bio-availability, preventing enteric pathogens colonization, restoring gut microflora, lowering pH and improving mucosal immunity. Therefore, probiotics can act as an ideal candidate to improve the general health and productivity of ruminants (Sharma *et al.*, 2018). Continuous antibiotic treatment usually results in washing out of indigenous gut flora, which can be replenished with the concomitant administration of selective probiotic and prebiotic components. Probiotics display direct antagonistic activity against different pathogens and multi-drug resistant strains such as, *Pseudomonas aeruginosa*, *Acinetobacter baumannii*, *Salmonella* and *Shigella* spp. *etc.* In this line, *Lactobacillus acidophilus* was shown to display strong antagonistic activity against MDR *P. Aeruginosa* (Jamalifar *et al.*, 2011). *Bifidobacterium pseudocatenulatum* SPM1309 was reported to inhibit MDR *P. Aeruginosa* and *A. Baumannii* (Lee *et al.*, 2013). *L. casei* strain exhibited strong activities against MDR *Shigella* strain (Mirnejad *et al.*, 2013). Keeping in mind the strong antagonistic activity of probiotic strains, several strains such as, *Lactobacillus*, *Bacillus*, *Enterococcus*, *Bifidobacterium* and *Saccharomyces* spp. *etc.* are used as probiotic in livestock and animal husbandry (Sharma *et al.*, 2018). Prebiotics support selective proliferation of probiotic bacteria and also display anti-microbial activities and promote immune system. The commonly used prebiotics are oligosaccharides, polysaccharides, polyols and protein hydrolysates. Celmanax, a commercial prebiotic for cattle prevent enterohemorrhagic *E. coli* colonization in the intestine (Baines *et al.*, 2011). In an *in-vivo* study, continuous oral co-administration of *Bifidobacterium breve* and galacto-oligosaccharide, a prebiotic rendered protection against MDR *A. Baumannii* and increased the survival rates of mice (Asahara *et al.*, 2016). The application of pro-, pre- and syn- biotics in dairy animals is currently limited due to paucity of data defining the strain specific activity of probiotic strains; variability within host animal species; age, diet and physical condition of target animal; degree of the pathogen challenge and other stressors; dose and duration of administration etc. All these factors need to be considered and standardized.

7.2 Bacteriocins

Bacteriocins are the ribosome synthesized bacterial antimicrobial peptides that can kill or inhibit closely-related bacterial strains. Most commonly, lactic acid bacteria (LAB) are known to secrete bacteriocins *i.e.* nisin (*Lactococcuslactis*), lactocin (*Lactobacillus sakei*), pediocin (*Pediococcusacidi lactici*), acidocin (*L. acidophilus*), sacacin (*L. sakei*), plantaricin (*L. plantarum*), helveticin (*L. helveticus*), curvacin (*L. curvatus*), lactobin (*L. amylovorus*) *etc.* Bacteriocins produced by gram-negative bacteria are colicins and microcins (*E. coli*). Among LAB bacteriocins, nisin is the most commonly explored. Bacteriocins have wide application in food preservation and antimicrobial packaging. Research studies have shown that co-application of bacteriocins and antibiotics significantly improve the effectiveness of drugs. Addition of nisin along with standard doses of antibiotics reduced the MIC values of all the tested drugs against *E. faecalis* (Tong *et al.*, 2014). Bacteriocins based products have found several applications in veterinary and human medicine. Different bacteriocin based products *i.e.* teat sanitizers, teat dips, dairy wipes are being used for management of mastitis in diary animals and have been reported to be effective against *S. aureus*, *S. dysgalactiae* and *S. uberis* (Pieterse and Todorov 2010). Several bacteriocins *i.e.* Geobacillus I and Geobacillus II, nisin U, nisin Z, uberolysin, bacteriocin ST91KM, morricin, kurstacin, kenyacin, entomocin, Pep5, epidermin, epilancin, epicidin, aureocins, microcin, colicins, enterocin *etc.* have displayed potent antimicrobial activity against pathogens. Enterocin (Sharma *et al.*, 2018). Therefore, application of bacteriocins as co-therapy can reduce dependency on antibiotics.

7.3 Antimicrobial Peptides (AMPs)

Antimicrobial Peptides (AMPs) are one among the leading alternatives to ongoing antibiotic therapy. AMPs display remarkable immune-modulator and broad-spectrum antimicrobial against bacteria and fungi, which make them an ideal candidate for the development of novel therapeutics. These amphipathic polypeptides intercalate into the bacterial membrane, creating pores and resulting in osmotic lysis (see Figure 1).

Additionally, AMPs bypass the common resistance mechanisms employed by drug resistant pathogens. Ruminants possess a wide spectrum of AMPs that offers innate defence against microbial pathogens (Brogden *et al.*, 2003). Islas-Rodriguez *et al.* (2009) reported potent and rapid activity of AMP Esc-1a (1–21) NH_2 against bovine mastitis pathogen, *Streptococcus agalactiae*. Bacitracin methylene salicylic acid and bacitracin zinc had been approved in USA and China as feed additives (Cheng *et al.*, 2014). AMPs have also been screened for their

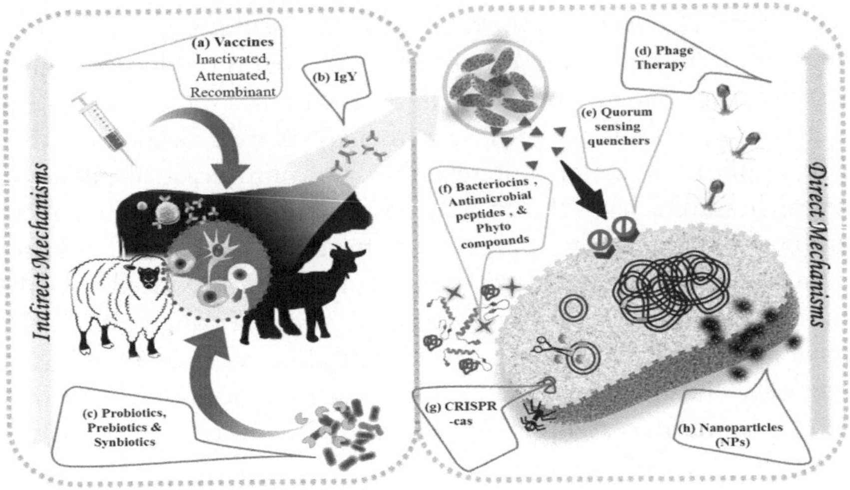

Figure 1: Mode of Action of Different Antimicrobial Resistance Management Strategies (adapted from Sharma *et al.*, 2018)

antimicrobial potency against pathogens exhibiting multi-drug resistance. Beside AMPs, natural milk proteins also act as source for many biologically active antimicrobial peptides. This milk derived AMPs have found application as both dietary supplements and as drugs (Akalin *et al.*, 2014). Theolier and co-workers (2014) established a milk AMP database containing microbiological and physicochemical data on AMPs. The database comprised of 371 entries in 2014 that has increased to 944 entries in 2017 (Nielsen *et al.*, 2017). Earlier, Mandal *et al.* (2014) reported broad spectrum antimicrobial activities of lactoferrin derived peptide and kappa casein short chain peptides. With the discovery of natural and synthetic AMPs, attention is being extended to structurally modified AMPs. Earlier, Lee and co-workers (2014) designed a helix-PXXP-helix (HPA3P2) structure that increased the antibacterial activity against multidrug resistant *P. aeruginosa* in mice model. Collectively, these findings indicate that the modification of peptides can enhance their antibacterial activity manifold and can be explored as an effective therapy against MDR strains. Peptide engineering has fuelled the discovery of short and potent AMP derivatives, several of which are currently evaluated under clinical trials (Padhi *et al.*, 2014). In another study, L10 (WFRKQLKW), a modified peptide from bovine lactoferrin displayed potent antibacterial and antifungal activity against ESBL producing gram-negative bacteria and *Candida* isolates. AMPs possess potent bactericidal effects and are easily metabolized. Besides, there is growing focus on the application of vaccines, immunoglobulin, phages, photochemical, quorum quenchers, nanoparticles and CRISPR cas as alternatives and/or co-therapy to antibiotic resistance; however, their application in food matrix is questionable due to sensory and overall food quality issues.

7.4 Dynamic AMPs

Recently, synthetic antimicrobial polymers have emerged as alternatives to natural AMPs, and it has been shown that multivalency (*i.e.* multi polymer chains in the same entity) leads to more potent antimicrobials with less off-target toxicity, minimizing the safety concerns with natural AMPs. For example, synthetic star-shaped AMP nanoparticles (NPs) that have multiple polypeptide arms have been shown to be highly potent *in-vitro* and *in-vivo* towards gram-negative bacteria and do not contribute towards antimicrobial resistance as they act through multiple mechanisms (Lam *et al.,* 2016). In addition to multivalency, another promising concept that could further enhance the potency and reduce resistance development of synthetic AMPs is that of dynamic functionalities. This concept has been demonstrated for ligands that are free to move along a polymer backbone, maximizing receptor interactions and binding affinities (Sharma *et al.,* 2018). Ideal alternatives to current day antibiotics should be stable, should have a high antimicrobial activity at minimized doses, besides being non-toxic to the host, and above all free from resistance. The antibiotic alternatives should have minimum or no effect over resident gut flora.

8. CONCLUSIONS

Food, primarily of animal origin can serve as a vehicle for transmission of resistant strains and antimicrobial resistance traits. Further, processing conditions also augments the resistance profile of microbial strains. Globalization and establishment of a global market has further augmented the free mobility of antimicrobial resistant strains and probability of food crises and incidents. Thus, there is an urgent need for a global response plan to design mitigation strategies and practices for control of multidrug resistance strains and their transmission thereof by the implementation of an effective system of food risk intelligence on a global scale.

REFERENCES

[1] Akalın, A.S. (2014). Dairy-derived antimicrobial peptides: Action mechanisms, pharmaceutical uses and production proposals. *Trends in Food Science & Technology* 36: 79–95.

[2] Ameh, E.M., Tyrrel, S., Harris, J., Ignatiou, A., Orlova, E. and Nocker, A. (2018). The absence or presence of a lytic coliphage affects the response of *Escherichia coli* to heat, chlorine, or UV exposure. *Folia Microbiologica,* 63:599–606.

[3] Asahara, T., Takahashi, A., Yuki, N., Kaji, R., Takahashi, T. and Nomoto, K. (2016). Protective effect of synbiotic against multidrug-resistant *Acinetobacterbaumannii* in a murine infection model. *Antimicrobial Agents and Chemotherapy* 60: 3041–3050.

[4] Baines, D., Erb, S., Lowe, R., Turkington, K., Sabau, E., Kuldau, G., Juba, J., Masson L. Mazza, A. and Roberts, R. (2011). A prebiotic, Celmanax™, decreases *Escherichia coli* O157: H7 colonization of bovine cells and feed-associated cytotoxicity *in-vitro*. *BMC Research Notes,* 4:110.

[5] Borriello, S.P. (1986). Microbial flora of the gastrointestinal tract. *Microbial metabolism in the digestive tract,* pp. 2–16.

[6] Brogden, K.A., Ackermann, M., McCray Jr., P.B. and Tack, B.F. (2003). Antimicrobial peptides in animals and their role in host defences. *International journal of antimicrobial agents,* 22: 465–478.

[7] Cheng, G., Hao, H., Xie, S., Wang, X., Dai, M., Huang, L. and Yuan, Z. (2014). Antibiotic alternatives: the substitution of antibiotics in animal husbandry? *Frontiers in Microbiology,* 5: 217.

[8] Grasselli, E., François, P., Gutacker, M., Gettler, B., Benagli, C., Convert, M., Boerlin, P., Schrenzel, J. and Piffaretti, J.C. (2008). Evidence of horizontal gene transfer between human and animal commensal *Escherichia coli* strains identified by microarray. *FEMS Immunology and Medical Microbiology,* 53: 351–358.

[9] Guillermo, T., Andrea, L., Juan, D.L., Xochitl, H.V., Billy, M.H. and Todd, C. (2015). Food-producing animals and their health in relation to human health. *Microbial Ecology in Health and Disease,* 26: 25876.

[10] Hanselman, B.A., Kruth, S.A., Rousseau, J., Low, D.E., Willey, B.M., McGeer, A. and Weese, J.S. (2006). Methicillin-resistant *Staphylococcus aureus* colonization in veterinary personnel. *Emerging infectious diseases* 12: 1933.

[11] Islas-Rodrìguez, A.E., Marcellini, L., Orioni, B., Barra, D., Stella, L. and Mangoni, M.L. (2009). Esculentin 1–21: a linear antimicrobial peptide from frog skin with inhibitory effect on bovine mastitis-causing bacteria. *Journal of peptide science: an official publication of the European Peptide Society* 15: 607–614.

[12] Jamalifar, H., Rahimi, H.R., Samadi, N., Shahverdi, A.R., Sharifian, Z., Hosseini, F., Eslahi, H. and Fazeli, M.R. (2011). Antimicrobial activity of

different *Lactobacillus* species against multi-drug resistant clinical isolates of *Pseudomonasaeruginosa. Iranian Journal of Microbiology,* 3: 21.

[13] Lam, S.J., O'Brien-Simpson, N.M., Pantarat, N., Sulistio, A., Wong, E.H., Chen, Y.Y., Lenzo, J.C., Holden, J.A., Blencowe, A., Reynolds, E.C. and Qiao, G.G. (2016). Combating multidrug-resistant Gram-negative bacteria with structurally nanoengineered antimicrobial peptide polymers. *Nature Microbiology,* 1: 16162.

[14] Lee, D.K., Kim, M.J., Kang, J.Y., Park, J.E., Shin, H.S. and Ha, N.J. (2013). Probiotics with antimicrobial activity against multidrug resistant *Pseudomonas aeruginosa* and *Acinetobacterbaumannii. The Korean Journal of Microbiology,* 49: 245–252.

[15] Lee, J.K., Park, S.C., Hahm, K.S. and Park, Y. (2014). A helix-PXXP-helix peptide with antibacterial activity without cytotoxicity against MDRPA-infected mice. Biomaterials, 35: 1025–1039.

[16] Lewis, H.C., Mølbak, K., Reese, C., Aarestrup, F.M., Selchau, M., Sørum, M. and Skov, R.L. (2008). Pigs as source of methicillin-resistant *Staphylococcus aureus* CC398 infections in humans, Denmark. *Emerging infectious diseases,* 14: 1383

[17] Lou, Y. and Yousef, A.E. (1997). Adaptation to sublethal environmental stresses protects Listeria monocytogenes against lethal preservation factors. *Applied and Environmental Microbiology,* 63: 1252–1255.

[18] Mandal, S.M., Bharti, R., Porto, W.F., Gauri, S.S., Mandal, M., Franco, O.L. and Ghosh, A.K. (2014). Identification of multifunctional peptides from human milk. *Peptides,* 56: 84–93.

[19] Mirnejad, R., Vahdati, A.R., Rashidiani, J., Erfani, M. and Piranfar, V. (2013). The antimicrobial effect of *Lactobacillus casei* culture supernatant against multiple drug resistant clinical isolates of *Shigellasonnei* and *Shigellaflexneri in-vitro. Iranian Red Crescent Medical Journal,* 15: 122.

[20] Muñoz-Atienza, E., Gómez-Sala, B., Araújo, C., Campanero, C., Del Campo, R., Hernández, P.E., Herranz, C. and Cintas, L.M. (2013). Antimicrobial activity, antibiotic susceptibility and virulence factors of lactic acid bacteria of aquatic origin intended for use as probiotics in aquaculture. *BMC Microbiology,* 13:15.

[21] Nielsen, S.D., Beverly, R.L., Qu, Y. and Dallas, D.C. (2017). Milk bioactive peptide database: A comprehensive database of milk protein-derived bioactive peptides and novel visualization. *Food Chemistry,* 232: 673–682.

[22] Padhi, A., Sengupta, M., Sengupta, S., Roehm, K.H. and Sonawane, A. (2014). Antimicrobial peptides and proteins in mycobacterial therapy: current status and future prospects. *Tuberculosis,* 94: 363–373.

[23] Pieterse, R., Todorov, S.D. and Dicks, L.M. (2010). Mode of action and *in-vitro* susceptibility of mastitis pathogens to macedocin ST91KM and preparation of a teat seal containing the bacteriocin. *Brazilian Journal of Microbiology,* 41: 133–145.

[24] Pleitner, A., Zhai, Y., Winter, R., Ruan, L., McMullen, L.M., Ganzle, M.G. (2010). Compatible solutes contribute to heat resistance and ribosome stability in *Escherichia coli* AW1.7. *Biochimica et Biophysica Acta* 1824:1351–1357.

[25] Sharma, C., Rokana, N., Chandra, M., Singh, B.P., Gulhane, R.D., Gill, J.P.S., Ray, P., Puniya, A.K. and Panwar, H. (2018). Antimicrobial resistance: its surveillance, impact, and alternative management strategies in dairy animals. *Frontiers in Veterinary Science,* 4: 237.

[26] Théolier, J., Fliss, I., Jean, J. and Hammami, R. (2014). Milk AMP: A comprehensive database of antimicrobial peptides of dairy origin. *Dairy Science and Technology,* 94: 181–193.

[27] Tong, Z., Zhang, Y., Ling, J., Ma, J., Huang, L. and Zhang, L. (2014). An *in-vitro* study on the effects of nisin on the antibacterial activities of 18 antibiotics against *Enterococcus faecalis*. *PLoS One,* 9:e89209.

[28] Yang, Y., Kadim, M.I., Khoo, W.J., Zheng, Q., Setyawati, M.I., Shin, Y.J., Lee, S.C. and Yuk, H.G. (2014). Membrane lipid composition and stress/virulence related gene expression of *Salmonella enteritidis* cells adapted to lactic acid and trisodium phosphate and their resistance to lethal heat and acid stress. *International Journal of Food Microbiology,* 191: 24–31.

Antibiotic Residues in Food of Animal Origin

Jasbir Singh Bedi*, **Prateek Jindal and R.S. Aulakh**
School of Public Health and Zoonoses
Guru Angad Dev Veterinary and Animal Sciences University (GADVASU), Ludhiana (Punjab), India
E-mail: *bedijasbir78@gmail.com

ABSTRACT: Antibiotics are widely used in food animals to treat, control and prevent infections as well as for promotion of growth and production efficiency. The use of antibiotics in food-producing animals may result in occurrence of their residues in milk, meat, eggs, honey etc. The consumption of food or food products containing antibiotic residues may pose potential health risk for consumers including allergic reactions, disorders of intestinal flora, mutagenicity, nephropathy, hepatotoxicity, reproductive disorders etc. In addition to these, the other important aspect related to excessive use of antibiotics in humans and veterinary medicine is the emergence of resistant pathogens. The occurrence of residues in foods may possibly occur due to negligence regarding antibiotic withholding periods following use in animals, dry cow intra-mammary therapy, extra-label dosages for animals, or the use of prohibited antibiotics in animals. Various analytical methods are in practice to identify and quantify antibiotic residues in food such as ELISA, microbial inhibition, HPLC, GC-MS and LC-MS with varying degree of sensitivity and specificity. It is necessary to establish sensitive and reliable analytical methods for the determination of residues of different classes of antimicrobials in food. Moreover, in view of the consumer safety, importance to dairy industry and emergence of antimicrobial resistance, there is a need to establish surveillance systems for obtaining reliable data on veterinary antimicrobial usage and their residues in food of animal origin in India.

Keywords: Antibiotics, Residues, Animal Food, Detection Methods, Regulations, India.

1. INTRODUCTION

Antibiotics are the agents that have either bactericidal (kill) or bacteriostatic (inhibit growth) effect on bacteria. Antibiotics are produced by microorganisms as well as synthetically and are used for human, animal and plant health. Veterinary drugs/antibiotics are mainly used for therapeutic, prophylactic, diagnostic purposes or modification of any physiological function or behavior in animals. The classification of antibacterial antibiotics is usually done on the basis of their mechanism of action, chemical structure or spectrum of activity. Majority of antibacterial targets the growth or functions of bacteria. Bactericidal activities are shown by antibiotics which target the bacterial cell wall, cell membrane or interfere with essential bacterial enzymes whereas bacteriostatic activities are shown by antibiotics targeting bacterial protein synthesis. Antibiotics can be further classified as narrow and broad spectrum antibiotics based on their target specificity.

Table 1: Classification of Antibiotics based on Mechanism of Action

Mechanism of Action	*Antibiotic*	*Action*
Target bacterial cell wall	Penicillins, Cephalosporins	Bactericidal
Target bacterial cell membrane	Polymyxins	Bactericidal
Interfere with essential bacterial enzymes	Rifamycins, Lipiarmycins, Quinolones, Sulphonamides	Bactericidal
Target protein synthesis	Aminoglycosides, Macrolides, Tetracyclines	Bacteriostatic

The growing usage of antibiotics in food producing animals for prophylactic and therapeutic purposes and to enhance growth and productivity has led to the strengthening of animal husbandry sector in India. The increasing food demand for expanding human population has led to excess use of antibiotics in livestock sector. India has emerged as the world's popping antibiotic capital with a 62% hike in the use of antibiotics in the last decade. A 36% rise has been seen worldwide in the use of antibiotics as per the report "Global Trends in Antibiotic Consumption" from 2000–2010, of which 27% is a result of BRICS countries—Brazil, Russia, India, China and South Africa. In the present scenario, antibiotic residues are being detected in foods of animal origin such as milk, eggs, meat etc. which could be related to injudicious use or non-compliance of withdrawal period of antibiotics in food producing animals. These residues have very harmful health effects in humans and also lead to the emergence of antibiotic resistant pathogens. The use of antimicrobial drugs in excess amounts could result in deposition of antimicrobial residues in animals. Consumption of these residues in animal products may cause health risk to consumers including development of antibiotic resistance, carcinogenic effects, disruptions of normal intestinal flora, mutagenic effects, and hypersensitivity reactions. The growing number of animals raised for food production has resulted in sixty-seven percent of worldwide increase in antimicrobial use, which agrees with the fact that global consumption of antibiotics in animals is twice that of humans (Van Boeckel *et al.,* 2015; Aarestrup 2012). Global consumption of antimicrobials in food animal production was estimated to be 63,151 tons in 2010 and is projected to rise by 67%, to 105,596 tons, by 2030.

Antibiotic residue in food is defined as the amount of antibiotics at or above the tolerance or safe concentration for approved drugs, or at any concentration for unapproved drugs or at any concentration for illegal drugs. The food producing animals may get exposed to antimicrobials by many direct and indirect routes. The direct route involves addition of low doses of antimicrobials to animal feed to act as animal growth promoters for improving feed conversion efficiency and increasing lean to fat ratio. Further, antibiotics are also used in prophylactic and therapeutic measures in animals. It has been observed that after administration of antibiotics, 30–90% of the given dose is either excreted as non-metabolized or active metabolites forms (Pinheiro *et al.,* 2013) resulting into high levels of antibiotics and active metabolites to be present in faeces or urine (Masse *et al.,* 2014). Sometimes the antibiotic metabolites can transform back into their parent compound after excretion (Griffiths *et al.,* 1999).

The food producing animals may also get exposed through indirect routes that involve contaminated environment such as effluents from antimicrobial production plants and waste of animals and humans treated with antibiotics. The sewage from these sources when without being treated or after partial treatment released into the water bodies contaminate the surface and ground water. Such water when consumed by food producing animals exposes them to very low doses of antibiotics. After the application on animal waste on land, the active components may travel to soil or surface/ground water and pose detrimental impacts after contact with humans as well as environment. (Boxall *et al.*, 2003).

There are many reasons for occurrence of drug residues in food of animal origin particularly milk, meat and eggs (Figure 1). Residues may appear in food because of non-follow up of withdrawal periods for antibiotics used in animals, extended usage of antibiotics, multiple dosing (more than recommended), extra label use, poor record keeping of animals, etc. Sometimes dairy farmers contaminate the milk by mixing milk from treated animals with non-treated animal's milk.

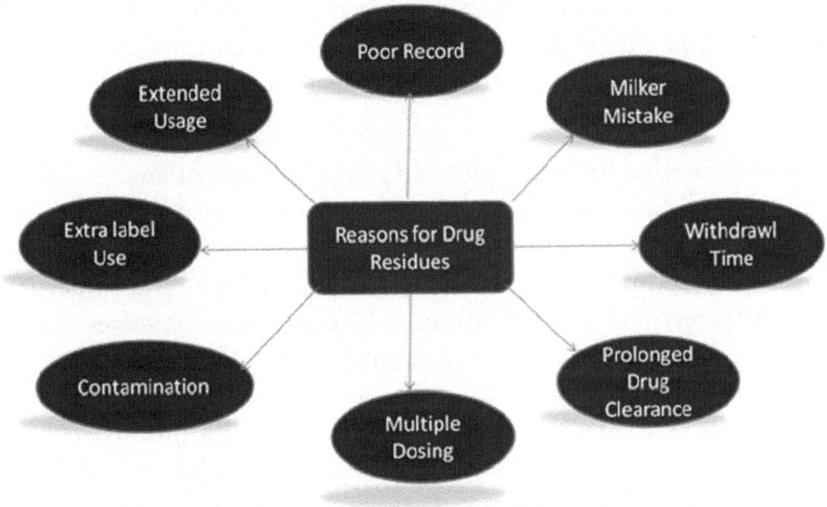

Figure 1: Reasons for Drug Residues in Food of Animal Origin

2. OCCURRENCE OF ANTIBIOTIC RESIDUES IN FOODS OF ANIMAL ORIGIN IN INDIA

Antibiotics are being used in a widespread manner in food producing animals which results in the presence of drug residues in foods of animal origin such as meat, milk and eggs as well as in environment through fecal and urinary excretions. The occurrence of these residues in milk occurs primarily due to negligence regarding antibiotic withholding periods of milk following use of intra-mammary preparations in lactating cows which are observed in many developed countries. In addition, extra-label dosages for animals may also contribute. In case of meat animals, inappropriate withholding times after antibiotic injecttions can lead to occurrence of antibiotic residues in meat following

Table 2: Antibiotic Residues in Food of Animal Origin in India

Type of Food Sample	*State*	*Antibiotics Reported*	*Concentration of Antibiotics*	*Reference*
Milk (pooled and individual animal milk samples)	Hyderabad	Oxytetracycline residues • 9% pooled animal milk samples • 73% individual animal milk samples	• 0.2 to 1.4 µg/ml in pooled • 0.2 to 6.7 µg/ml in individual animal milk	Sudershan and Bhat 1995
Milk (Small dairy farm)	Punjab	• Enrofloxacin (13/164) 7.9% • Oxytetracycline (6/164) 3.7% • Tetracycline (5/164) 3.0% • Sulphamethoxazole (3/164) 1.8%	• Enrofloxacin 64.6–213.1 ng/ml • Oxytetracycline 55.5–177 ng/ml • Tetracycline 60.1–149 ng/ml • Sulphamethazole 71.2–93.4 ng/ml	Pallavi *et al.*, 2016
Milk (Market milk)	Punjab	• Enrofloxacin (3/32) 9.4% • Oxytetracycline (1/32) 3.1%	• Enrofloxacin 64.7–131.3 ng/ml • Oxytetracycline 77 ng/ml	Pallavi *et al.*, 2016
Milk	Punjab	Tetracycline (18/133)	16–134.5 ppb	Gaurav 2015
Milk	Bihar	• Tetracycline • Oxytetracycline • Sulfadimidine • Sulfamethoxazole	Out of 250 samples, 5, 2, 2 and 1 milk sample exceeded MRLs for tetracycline, oxytetracycline, sulfadimidine and sulfamethoxazole	Nirala *et al.*, 2017
Buffalo meat	Different parts of India	Oxytetracycline 5/122 (4.1%)	Lower than MRL	Biswas *et al.*, 2007

(Table contd...)

(Table 2 contd...)

Type of Food Sample	*State*	*Antibiotics Reported*	*Concentration of Antibiotics*	*Reference*
Chicken	Delhi NCR	28/70 (40%) showed the presence of antibiotic residues. • Tetracycline (oxytetracycline, chlortetracycline and doxycycline) 10/70 (14.3%) • Fluoroquinolones (enrofloxacin and ciprofloxacin) 20/70 (28.6%)	• Tetracycline 16.01–46.02 µg/kg • Fluoroquinolones 3.37–196.34 µg/kg	Sahu and Saxena 2014
Shrimps	Kerala, Tamil Nadu, Karnataka and Andhra Pradesh	Chloramphenicol, Sulphonamide, Erythromycin	• Chloramphenicol Trace level <1 ppb • Sulphonamide, Erythromycin <100 ppb	Swapna K.M. *et al.*, 2012
Eggs (from poultry farms)	Punjab	• Oxytetracycline (7/81) 8.6% • Chlortetracycline (4/81) 4.9% • Enrofloxacin (4/81) 4.9% • Ciprofloxacin (2/81) 2.5%	• Oxytetracycline 17.2 ± 75.4 µg/kg • Two samples for oxytetracycline exceeded MRL • Chlortetracycline 4.3 ± 18.6 µg/kg • Enrofloxacin 3.2 ± 13.3 µg/kg • Ciprofloxacin 1.4 ± 9.2 µg/kg	Lalawmpuia *et al.*, 2015

slaughter. In poultry, the use of antibiotics for treatment or growth promotion can result in residues in eggs. The consumption of food or food products containing antibiotic residues may present potential health risks for consumers. These risks include allergic reactions, disruption of intestinal flora, immunopathological effects, carcinogenicity, mutagenicity, nephropathy, hepatotoxicity, reproductive disorders, and bone marrow depression. Also, the presence of antibiotic residues in milk may interfere with starter culture for cheese and other dairy products resulting in significant economic losses to producers and manufacturers of milk and milk products. The knowledge, attitudes and practices of farmers regarding the use of antimicrobials in animal husbandry management practices at farm level influence the presence of drug residues in the final product. The presence of veterinary drug residues in food has become an issue of intense international concern, therefore the Joint FAO/WHO Expert Committee of Food Additives and Contaminants (JECFA)/Codex Alimentarius Commission (CAC), at its 35th Session of the CAC (July 2012) updated the Maximum Residue Limits (MRLs) for veterinary drugs in food. In India, published reports on the occurrence of antibiotic residues in food are scarce possibly due to lack of sufficient expertise and resources in veterinary institutes and non-existence of national limits for antibiotic residues in foods of animal origin.

3. REGULATIONS FOR USE OF ANTIBIOTICS IN FOOD PRODUCING ANIMALS IN INDIA

In 2013, the first amendment was made in Drugs and Cosmetics Rules 1945 regarding authorized guidelines on the use of antibiotics in food animals reared for domestic consumption by the Ministry of Health and Family Welfare, Government of India. According to the amendment, the withdrawal period of the concerned medicine used in treatment of food-producing animals must be mentioned on the container. The recommended withdrawal period for different commodities is as mentioned i.e. at least 7 days for eggs or milk, 28 days for meat including fat and offal and 50 days for fish flesh. In India, the class-prescription drugs under Schedule H have been recently introduced which can only be procured with the prescription of a licensed doctor. The various standards in the food commodities are regulated by Food Safety and Standards Authority of India (FSSAI) by the

Ministry of Health and Family Welfare. Currently the maximum tolerance levels for antibiotics and other pharmacologically active substances have been established under the Food Safety and Standards (contaminants, toxins and residues) Regulations, 2011 by FSSAI for variety of sea foods, fish and fishery products only. However, till date no tolerance limits have been set for antibiotics and other pharmacologically active substances in milk, meat and eggs.

Table 3: The Codex Maximum Residues Limits (MRLs) of Veterinary Drugs in Milk

Antibiotic	*MRL (μg/kg)*
Benzylpenicillin	4
Ceftiofur	100
Chlortetracycline, oxytetracycline and tetracycline	100
Cloxacillin	30
Dihydrostreptomycin and streptomycin	200
Enrofloxacin and Ciprofloxacin	100
Gentamycin	200
Neomycin	1500
Spectinomycin	200

Source: Codex Alimentarius Commission, 2012. Maximum residue limits for veterinary drugs in foods updated as at the 35th session.

4. PREVENTION OF ANTIBIOTIC RESIDUES FROM FOOD PRODUCING ANIMALS

Since antibiotics are used in animal husbandry for therapy and prophylaxis as well as to increase growth and feed efficiencies, the control of antibiotics in animals and foodstuffs of animal origin is indispensable. In recent years, the presence of veterinary drug residues in food has become an international concern. Avoidance of antibiotic residues in food is an important focus of many regulatory agencies. Limiting the availability of antimicrobials, enhanced surveillance and on-farm interventions are key strategies to reduce antibiotic residues in food animals. Improved, rapid diagnostic methods and accelerated development and approval of new antimicrobial drugs can also play an important role in preventing antimicrobial resistance. Antibiotic residues in foods and antimicrobial resistance can be prevented by:

- Prevention of diseases in food animals to minimize the use of antibiotics by maintenance of good herd health, by employing best management practices, early and accurate diagnosis of sick animals, implementing farm and herd biosecurity measures and timely vaccination of all animals in herd.
- Follow label directions for approved drugs (treatment regimen and withdrawal times).
- Extra-label use of antibiotic only under veterinarian guidance.
- Identify treated animals and milk into separate bucket.
- Establish effective record-keeping.
- Education/communication with all personnel involved with milking and treating cows and heifers.
- Establish a valid veterinarian/client/patient relationship.
- Use antibiotic residue screening tests for testing treated cow's milk.
- Timely training of people engaged in animal husbandry practices and treatment of livestock.
- Nationwide monitoring and surveillance of antibiotic residues in foods and food products of animal origin to avoid unacceptable health risks to consumers.

5. CONCLUSIONS

It is well evident that antibiotic residues are present in food of animal origin in India, and may pose risk to consumers' health. Thus, there is a need of stricter regulations for the use of antimicrobial drugs in the animal food industry as well as inspection of food for drug residues prior to marketing. The presence of these residues can be prevented at producer level by educating farm producers and making them aware to follow best animal husbandry practices to prevent infection and follow up of withdrawal period.

REFERENCES

[1] Aarestrup, F. (2012). Sustainable farming: Get pigs off antibiotics. *Nature* 486: 465–66.

[2] Biswas, A.K., Rao, G.S., Kondaiah, N., Anjaneyulu, A.S.R., Mendiratta, S.K., Prasad, R. and Malik, J.K. (2007). A Simple Multi-residue Method

for Determination of Oxytetracycline, Tetracycline and Chlortetracycline in Export Buffalo Meat by HPLC-Photodiode Array Detector. *Journal of Food and Drug Analysis,* 15(2): 1–8.

[3] Boxall, A.B.A., Koplin, D., Halling-Sorensen, B. and Tolls, J. (2003). Are veterinary medicines causing environmental risks? *Environmental Science and Technology,* 37: 286–94.

[4] Lalawmpuia, C. (2015). 'Monitoring of Antibiotic residues in poultry feed, water and eggs and its public health significance.' M.V.Sc Thesis, Guru Angad Dev Veterinary and Animal Science University, Ludhiana, India.

[5] Codex Alimentarius Commission (2012). Maximum residue limits for veterinary drugs in foods updated as at the 35th session.

[6] Gaurav, A. (2015). Studies on antibiotic residues in food of animal origin. *Antimicrobial Resistance and Infection Control,* 4: 168.

[7] Griffiths, B.S., Ritz, K., Ebblewhite, N. and Dobson, G. (1999). Soil microbial community structure: effects of substrate loading rates. *Soil Biology and Biochemistry* 31: 145–53.

[8] Masse, D.I., CataSaady, N.M. and Gilbert, Y. (2014). Potential of biological processes to eliminate antibiotics in livestock manure: An overview. *Animals* 4: 146–63.

[9] Pallavi (2016). 'Multi-residue detection of antibiotics in milk and role of dairy farm managerial practices in occurrence of residues.' Ph.D. Thesis, Guru Angad Dev Veterinary and Animal Science University, Ludhiana, India.

[10] Pinherio, A., Albano R.M.R., Alves, T.C., Kaufmann, V. and D Silva, M.R. (2013). Veterinary antibiotics and hormones in water from application of pig slurry to soil. *Agricultural Water Management* 129: 1–8.

[11] Sahu, R. and Saxena, P. (2014). Antibiotics in chicken meat. Centre for Science and Environment 41, Tughlakabad Institutional Area, New Delhi, pp. 1–36. https://www.researchgate.net/publication/266208046_Antibiotics_in_Chicken_Meat

[12] Sudershan, R.V. and Bhatt, R.V. (1995). A survey on veterinary drug use and residues in milk in Hyderabad. *Food Additives and Contaminants* 12(5): 645–50.

[13] Swapna, K.M., Rajesh, R. and Lakshmanan, T.P. (2012). Incidence of antibiotic residues in farmed shrimps from the southern states of India. *Indian Journal of Geo-Marine Sciences,* 41(4): 344–47.

[14] Van Boeckel, T.P., Brower, C., Gilbert, M., Grenfell, B.T., Levin, S.A., Robinson, T.P. and Laxminarayan, R. (2015). Global trends in antimicrobial use in food animals. *Proceedings of the National Academy of Sciences, U.S.A.,* 112(18): 5649–54.

Pulse Field Gel Electrophoresis and its Applications

Randhir Singh

School of Public Health and Zoonosis, Guru Angad Dev Veterinary and Animal Sciences University (GADVASU), Ludhiana (Punjab), India

E-mail: sainirandhir74@gmail.com

ABSTRACT: Molecular epidemiological tools play pivotal role in disease control and prevention. The key to achieve that is to find or trace the origin of the disease causing agents. This is critical as clonal relationship has to be established between similar isolated strains from different routes of disease transmission pathways. At the same time it also important to rule out unrelated microorganisms. Typing technologies plays important role to accomplish these objectives. Among the various typing methods, pulse field gel electrophoresis (PFGE) is considered the gold standard until sequencing based methods become more affordable. PFGE is based on the restriction analysis of bacterial genomic DNA especially chromosomal DNA. This technique is different from conventional electrophoresis as it can separate genomic DNA greater than 20 kb. In PFGE, the electric current is pulsed in two directions alternatively to separate molecular fragments of varying sizes. Contour-clamped homogeneous electric field (CHEF) system is popular and widely used among different variants of PFGE. Other systems of PFGE and their modes of working are discussed in this paper. PFGE is widely applied in strain typing of microorganisms associated with food safety and antimicrobial resistance. It is also popular in food industry for quality control application. Genome mapping and cancer research are the other areas where it is used.

Keywords: PFGE, DNA Fragments, Typing, CHEF.

1. INTRODUCTION

Infectious bacterial diseases cause significant impact on human and animal well-being. The ability to trace the primary source of bacterial contamination is key to improving human and animal health. Some of the tools that are based on typing technologies can be used to connect sick people with the source of contamination. Although several methods are there to estimate the distribution of infectious agents and their origin, but it is also equally important to differentiate unrelated isolates from a disease outbreak condition. As such isolates have the ability to complicate or confound the epidemiological investigation (Zadoks and Schukken 2006). Broadly, disease causing bacterial agents can be characterized on the basis of biochemical characters, phenotypic traits, antimicrobial resistance profiling, serotyping or by genome based molecular typing methods. Molecular typing methods mainly rely on three important discrimination mechanisms, those that use restriction enzyme digestion of bacterial DNA for restriction pattern analysis, those based on amplification of specific genetic targets and those based on analysis of sequenced bacterial genome for identification of sequence polymorphism (Foley *et al.*, 2009). Among the restriction pattern based analysis of bacterial DNA, Pulse Field Gel Electrophoresis (PFGE) technique discriminates bacterial isolates/strains on the basis of restriction pattern of the chromosomal DNA generated using restriction enzyme digestion.

2. PULSE FIELD GEL ELECTROPHORESIS (PFGE)

This method uses rare cutting restriction enzymes for the digestion of chromosomal DNA embedded in agarose plug. Among the molecular typing methods, this is often considered as '***Gold Standard***', until sequence based methods become more affordable. The restriction enzyme digested bacterial DNA fragments in plug are separated in high quality agarose by alternating electric field. Conventional DNA electrophoresis can separate DNA fragments up to 20 kb, however fragments bigger than 20 kb in sample migrate together giving large fuzzy image at the top of the gel. Schwartz and co-worker (Schwartz *et al.*, 1982, Schwartz and Canter 1984) came up with a technique to overcome this problem. They alternated the electric field between distinctly positioned pairs of electrodes to reorient DNA fragments as large as many megabases to let them migrate at different rates through the gel.

3. MECHANISM OF PFGE

DNA fragments exhibit diverse forms of motion when put under electric field, which results in different rates of movement/migration of different sized DNA molecules. Experiments have shown that DNA molecule migrating through the gel is elongated in the direction of electric field strength (Bustamante *et al.,* 1990, Gurrieri *et al.,* 1996). On changing or switching the field of strength, DNA molecule begins to move in new direction through one of their end while still retaining their elongated conformations. If the angle between the alternating fields of strength is 90°, then any tail of the molecule may become the leading end. If the angle is less than 90°, the frontal part of the molecule becomes the leader. However, if the angle of the field of strength is more than 90°, the posterior part of the molecule becomes the leading end. In PFGE devices angle between the alternate fields of strength is more than 90°, therefore DNA molecule starts to move in the new direction by their posterior end with an initial backward motion proportional to their size (Figure 1) resulting in large DNA molecule staying behind the small ones. The instrument designed by Schwartz and Cantor (1984) had reorientation angle of 90°, when the field of strength was switched after certain interval. This system generated homogeneous and non-homogeneous fields due to distinct position of dot electrodes. This was designated as "Pulsed Field Gradient Electrophoresis". They considered the gradient to be the key for the resolution of DNA fragments in pulsed field. Therefore, they used electrode positioned in a manner generating non-homogenous filed. In this electrophoresis system, angle between the fields of strength varied in different regions of the gel (110–150°). However there was problem in comparison of the DNA fragments run on the adjacent lanes making it difficult to estimate the size. Later on a number of other variants of electrophoresis in Pulsed field were designed that differed in positioning of electrodes and the angle of reorientation.

4. PFGE VARIANTS

4.1 Orthogonal Field Alternation Gel Electrophoresis (OFAGE)

This variant of PFGE was given by Carle and Olson (1984). In this, the electrodes were arranged symmetrically but were of unequal lengths (see Figure 1). The DNA fragments in this system migrated under two non-homogenous electric fields with reorientation angle in different regions of the gel varying from 120–150°. In OFAGE gels electrophoretic movement of the DNA fragments in central lanes remained straight allowing comparison of molecules migrating in neighbour lanes. However, electrophoretic mobility of the molecules in the outer lanes remained curved making analysis difficult.

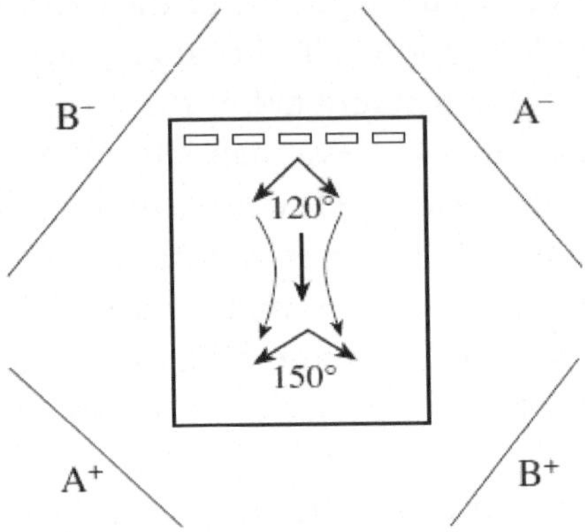

Figure 1: Schematic Drawing of OFAGE System (Nassonova, 2008)

4.2 Transverse Alternating Field Gel Electrophoresis (TAFE)

This variant of PFGE was suggested by Gardiner and co-workers (Gardiner *et al.*, 1986). In this system gel is placed vertically in the electrophoresis chamber and the field of strength which is non-homogeneous is applied across and not along the gel. The four electrode set is placed in front and at the back of the gel (see Figure 2). ll lanes experienced similar effects and the bands remain straight (Steward *et al.*, 1988). The electric field reorientation angle varies from 115° (at the top of the gel) to 165° (at the bottom). This resulted in reduced

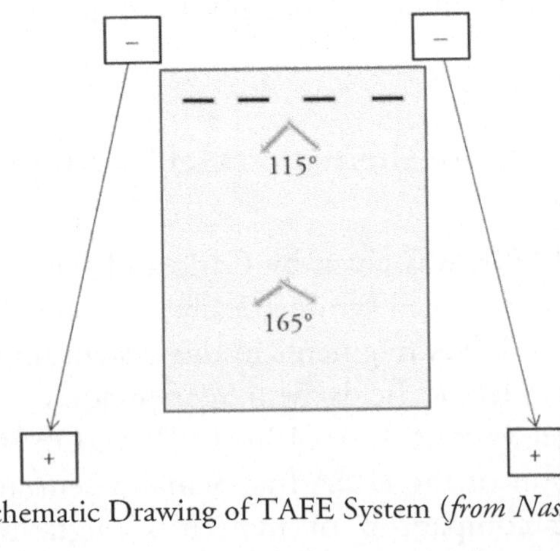

Figure 2: Schematic Drawing of TAFE System (*from Nassonova 2008*)

rate of DNA movement at the bottom of the gel, leading to band sharpening effect (Gardiner 1992). This problem made TAFE standardization and the result replication difficult. Despite the non-homogeneous electric field, the molecules in all lanes moved in straight and parallel lines, making comparison easy. This system can help separation of DNA bands up to 1600 kilo base.

4.3 Field Inversion Gel Electrophoresis (FIGE)

To overcome the problems of PFGE based on non-homogeneous electric field, technological progress were made for the newer PFGE systems. FIGE system was developed by Carle, Frank and Olson in 1986 (Carle *et al.,* 1986). In this system two electrodes of same length were placed parallel to each other in a horizontal chamber (see Figure 3). The field of strength is homogeneous across the gel and the polarity of the electric field is inverted periodically *i.e.* the reorientation angle of the migrating molecule is 180°. The net forward movement of the band was achieved due to difference in duration or in voltage strength during forward and reverse field application. FIGE is popular for small fragment separation. However, band inversion effect due to non-linear dependence of the migration rate of DNA from its size is seen.

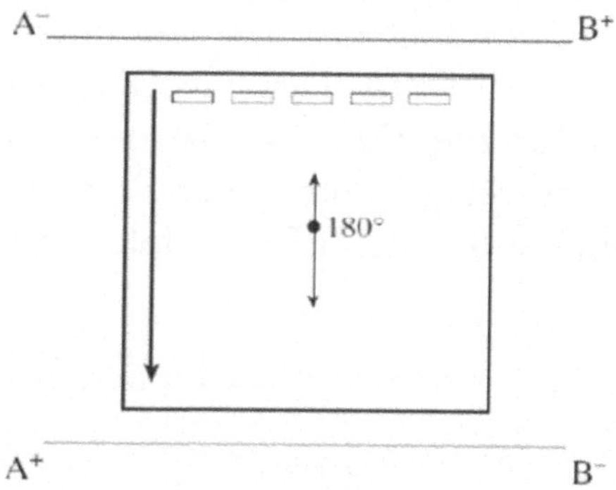

Figure 3: Schematic Drawing of FIGE System (*from Nassonova 2008*)

4.4 Contour-Clamped Homogeneous Electric Field (CHEF)

In 1986, Carle and co-workers (Carle *et al.,* 1986) came up with this variant of PFGE which is mostly widely used today. This is also based on application of homogeneous field throughout the gel. In this system, equally spaced twenty four electrodes are arranged in a hexa-gonal contour (see Figure 4). The DNA migration is independent of its position in the gel. This variant of the PFGE helps in stable DNA separation in straight lanes. The problem of band sharpening and band inversion is also not observed.

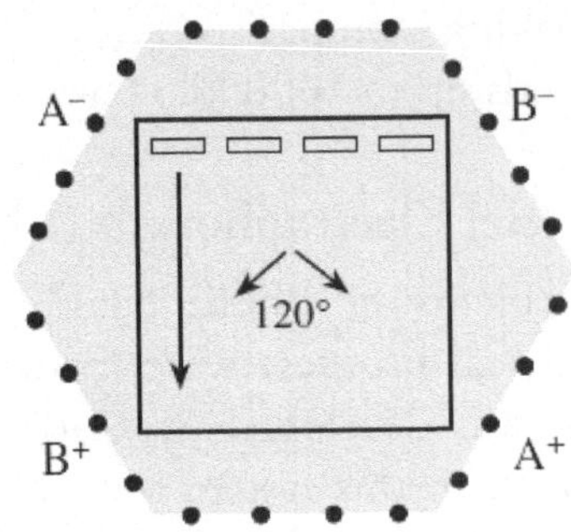

Figure 4: Schematic Drawing of CHEF System (*from Nassonova 2008*)

4.5 Rotating Gel Electrophoresis (RGE)

This system was described by Southern *et al.* (1987) in England. In this, pulsed field is generated not by switching the field of strength produced between the set of electrodes but by rotation of the gel on a horizontal platform at certain intervals and angle is called RGE (Figure 5). In another variant of this same system, the gel remains stationary but the electrodes move at certain angle and interval to generate field strength for separation of DNA fragments (Ziegler *et al.,* 1987). This system is named as ***Rotating Field Gel Electrophoresis (ROFE).*** Pulsed field electrophoresis working on other modifications are also described, however the system based on CHEF principle is most widely used.

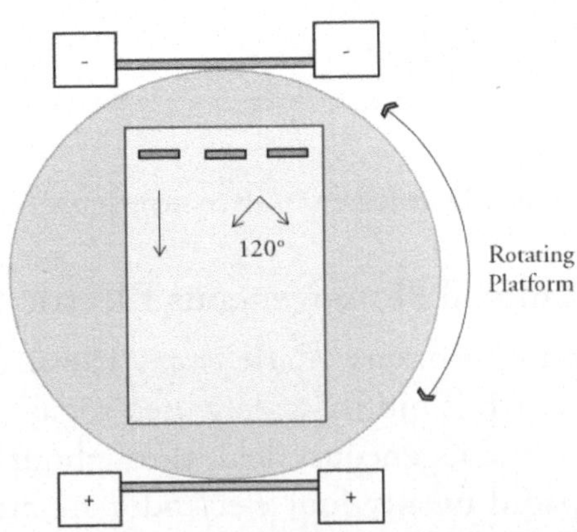

Figure 5: Schematic Drawing of RGE System (*from Nassonova 2008*)

5. FACTORS INFLUENCING PFGE

PFGE is a sensitive technique which is influenced by several factors native to the system that enable better separation of DNA fragments. Use of appropriate bacterial cell density, isolation of high quality pure DNA are however, the factors that are independent of the system and are key to better results.

5.1 Pulse Time (switch time, switch interval)

This is the length of time for which electric field is pulsed in single direction and then switched to another direction. Pulse time in pulsed electric field is critical for resolution of DNA molecules of different sizes. Selection of pulse time or switch time depends on the size of molecule. Larger the size of DNA fragments longer the pulse time should be. This is based on the assumption that optimal pulse time for molecular fragment is the time during which DNA fragments succeed in reorienting and starts to move in new direction. If the sample contains wide range of DNA fragments, then the fragments can be resolved by varying the pulse time or switch time over the period of run which is technically called as switch time ramping in which switch time is ramped from a short interval to a longer one. The switch time ramping can be linear or nonlinear in nature to achieve better separation of DNA fragments.

5.2 Pulse/Reorientation Angle

Pulse angle is important for achieving better resolution. Most of the current protocols use reorientation angle of 120°. It has been observed that change in the pulse angle did not affect migration of DNA fragments less than 1 Mb. Migration of DNA fragment increases with the decrease in angle, but reduces fragment resolution in gel (Clark *et al.,* 1988, Chu and Gunderson 1991). The best resolution is achieved with an angle more than 110° (Cantor *et al.,* 1988).

5.3 Voltage Gradient (Field Strength)

It is the strength of electric field per unit distance between the two electrodes and is represented by V/cm. In CHEF system the distance between the electrodes is approximately 33 cm, the voltage applied is

200 V which turns out to be approximately 6 V/cm. Separation of larger size DNA fragments (> 3 Mb) is better achieved while using low voltage gradients, 1.5–3 V/cm, coupled with narrow reorientation angle (Birren *et al.*, 1988). High voltage field strength results in poor separation and smearing effect (Chu 1990).

5.4 Buffer Type

Since the PFGE involves separation of large size DNA fragments, buffer with high ionic strength is used. Tris-borate-EDTA (0.5 × TBE) or Tris-acetate-EDTA (1 × TAE) are the two most commonly used buffers. Temperature of the buffer is another limiting factor that affects the migration of fragments through the gel. Generally, a buffer temperature of 12–15°C is the best for achieving better resolution of the bands. Increasing buffer temperature beyond this temperature range will reduce the run time but will also reduce the band resolution (Mathew *et al.*, 1988). On the contrary buffer temperature of 4°C will provide the sharper resolution, but will increase the run time significantly. Therefore, maintenance of buffer temperature permanently within above specified range is the key for good quality DNA resolution.

6. APPLICATIONS OF PFGE

PFGE is used in epidemiological studies of disease causing organisms of public health importance like *Escherichia coli* O157:H7, Campylobacter, *Vibrio cholera, Salmonella, Shigella, Listeria* (Chen *et al.*, 2004, Guard *et al.*, 2011, Swaminathan *et al.*, 2011, Vinas *et al.*, 2013) to identify variant strains and track food-borne pathogen source and spread. This technique is also helpful in comparing antimicrobial resistant strains of different organisms for clonal relationship and origin (Madiyarov *et al.*, 2010, White *et al.*, 2001). This is also used in food industry to study genetic stability of organisms for food quality control. Its application in genome mapping to identify rearrangements of genetic material for emergence of new strains or types, disease specific loci is also there. It is useful in generating chromosome libraries of prokaryotes and eukaryotes. Cancer research is another area where PFGE is used to study DNA damage and repair.

7. CONCLUSIONS

PFGE is an important epidemiological tool to study structural and functional analysis of genome in different organisms. CHEF is the most widely used PFGE system among the number of variants available. Mobility of the molecular fragments and its resolution is influenced by several interrelated parameters. The important ones are pulse time, field strength and reorientation angle. Efficient combination of all the factors based on targeted genome facilitates high quality resolution and separation of DNA molecules of different sizes. PFGE is such a vital molecular typing technique that it will continue to remain popular among research scientists generating enormous data for large number of users.

REFERENCES

[1] Birren, B.W., Lai, E., Clark, S.M., Hood, L. and Simon, M.I. (1988). Optimized conditions for pulsed field gel electrophoretic separations of DNA. *Nucleic Acids Res* 16:7563–7582.

[2] Bustamante, C., Gurrieri, S. and Smith, S.B. (1990). Observation of single DNA molecules during Pulsed-Field Gel Electrophoresis by Fluorescence Microscopy, in methods: A companion to methods in enzymology, San Diego: Academic, pp. 151–159.

[3] Cantor, C.R., Gaal, A. and Smith, C.L. (1988). High-resolution separation and accurate size determination in pulsed-field gel electrophoresis of DNA: Effect of electrical field shape. *Biochemistry,* 27:9216–21.

[4] Carle, G.F., Frank, M. and Olson, M.V. (1986). Electrophoretic separations of large DNA molecules by periodic inversion of the electric field. *Science* 232:65–68.

[5] Carle, G.F. and Olson, M.V. (1984). Separation of chromosomal DNA molecules from yeast by orthogonal-field-alternation gel electrophoresis. *Nucleic Acids Res.,* 12:5647–5664.

[6] Chen, C.H., Shimada, T., Elhadi, N., Radu, S. and Nishibuchi, M. (2004). Phenotypic and genotypic characteristics and epidemiological significance of ctx+ strains of *Vibrio cholerae* isolated from seafood in Malaysia. *Appl Environ Microbiol* 70:1964–1972.

[7] Chu, G. (1990). Pulsed-Field Gel Electrophoresis: Theory and Practice, in Methods: A Companion to Methods in Enzymology, San Diego: Academic Press, pp. 129–142.

[8] Chu, G. and Gunderson, K. (1991). Separation of large DNA by a variable-angle contour-clamped homogeneous electric field apparatus. *Anal Biochem* 194:439–446.

[9] Clark, S.M., Lai, E., Birren, B.W. and Hood, L. (1988). A novel instrument for separating large DNA molecules with pulsed homogeneous electric fields. *Science* 241:1203–1205.

[10] Foley, S.L., Lynne, A.M. and Nayak, R. (2009). Molecular typing methodologies for microbial source tracking and epidemiological investigations of Gram-negative bacterial foodborne pathogens. *Infect Genet Evol* 9: 430–440.

[11] Gardiner, K. (1992). Transverse Alternating-Field Electrophoresis, in Methods in Molecular Biology, Totowa: Humana Press, 12, pp. 51–61.

[12] Gardiner, K., Laas, W. and Patterson, D. (1986). Fractionation of large mammalian DNA restriction fragments using vertical pulsed-field gradient gel electrophoresis. *Somat Cell Mol Genet* 12:185–195.

[13] Guard, J., Morales, C.A., Fedorka-Cray, P., Gast, R.K. (2011). Single nucleotide polymorphisms that differentiate two subpopulations of Salmonella enteritidis within phage type. *BMC Res* 4:369-375.

[14] Gurrieri, S., Smith, S.B., Wells, K.S., Johnson, I.D. and Bustamante, C. (1996). Real-time imaging of the reorientation mechanisms of YOYO-labelled DNA molecules during 90 degrees and 120 degrees pulsed field gel electrophoresis. *Nucleic Acids Res* 24:4759–4767.

[15] Madiyarov, R.S., Bektemirov, A.M., Ibadova, G.A., Abdukhalilova, G.K., Khodiev, A.V., Bodhidatta, L. *et al.* (2010). Antimicrobial resistance patterns and prevalence of class 1 and 2 integrons in Shigella flexneri and Shigella sonnei isolated in Uzbekistan. *Gut Pathog* 2:18.

[16] Mathew, M.K., Smith, C.L. and Cantor, C.R. (1988). High-resolution separation and accurate size determination in pulsed-field gel electrophoresis of DNA. DNA size standards and the effect of agarose and temperature. *Biochemistry* 27: 9204–9210.

[17] Nasonova, E.S. (2008). Pulsed field gel electrophoresis: theory, instruments and applications. *Cell Tissue Biol* 2:557–565.

[18] Schwartz, D.C. and Cantor C.R. (1984). Separation of yeast chromosome-sized DNAs by pulsed field gradient gel electrophoresis. *Cell* 37: 67–75.

[19] Schwartz, D.C., Saffran W., Welsh J., Haas R., Goldenberg M. and Cantor C.R. (1982). New techniques for purifying large DNAs and studying their properties and packaging, cold spring harbor. *Symp Quant Biol.,* 47: 189–195.

[20] Southern, E.M., Anand, R., Brown, W.R. and Fletcher, D.S. (1987). A model for the separation of large DNA molecules by crossed field gel electrophoresis. *Nucleic Acids Res.,* 15: 5925–5943.

[21] Stewart, G., Furst, A. and Avdalovic, N. (1988). Transverse alternating field electrophoresis (TAFE). *Biotechniques,* 6: 68–73.

[22] Swaminathan, B., Barrett, T.J., Hunter, S.B. and Tauxe, R.V. (2001). CDC PulseNet Task Force. PulseNet: the molecular subtyping network for foodborne bacterial disease surveillance, United States. *Emerg Infect Dis.,* 7:382–389.

[23] Viñas, M.R., Tuduri, E., Galar, A., Yih, K., Pichel, M., Stelling, J. *et al.* (2013). Laboratory-based prospective surveillance for community outbreaks of *Shigella* spp. in Argentina. *PLoS Negl Trop Dis.,* 7: e2521.

[24] White, P.A., McIver, C.J. and Rawlinson, W.D. (2001). Integrons and gene cassettes in the enterobacteriaceae. *Antimicrob Agents Chemother,* 45: 2658–2661.

[25] Zadoks, R.N. and Schukken, Y.H. (2006). Use of molecular epidemiology in veterinary practice. *Vet Clin North Am Food Anim Pract,* 22: 229–261.

[26] Ziegler, A., Geiger, K.H., Ragoussis, J. and Szalay, G. (1987). A New electrophoresis apparatus for separating very large DNA molecules. *J. Clin Chem Clin Biochem,* 25: 578–579.

SECTION 2

Strategies for Mitigation of AMR

The Urgent Need for Safeguarding the Quality of Medicines in the Era of Antimicrobial Resistance: A Perspective

Chaitanya Kumar Koduri*, Phillip P. Nguyen and Kathrine Bond
United States Pharmacopeia, Hyderabad, India
E-mail: *ckk@usp.org

ABSTRACT: Antimicrobial Resistance (AMR) poses a significant threat to patients around the world by endangering our ability to treat common infections with the existing therapies. As part of the global call, India has developed an AMR National Action Plan that highlights concerted efforts to secure the quality of antimicrobials as a key strategy to better medicine stewardship, a core pillar of the global strategy against AMR. A focus on medicines quality could present a concrete path to tackling important contributors to AMR in ways that will also improve Indian health and regulatory system's capacities more generally.

Keywords: Antimicrobial Resistance (AMR), Public Health Policy, Drugs Quality; Quality Assurance, USP.

1. THE INTERSECTION OF AMR AND POOR QUALITY MEDICINES

The World Health Organization (WHO) calls AMR a "crisis that must be managed with the utmost urgency" and warns that "without urgent action we are heading for a post-antibiotic era, in which common infections and minor injuries can once again kill"—an "all-out" effort is needed. Globally, an estimated 700,000 people die each year from antimicrobial resistance.[1] By 2050, 10 million people may die from causes attributable to AMR.[2] Even in India, a study by The Lancet Infections Disease Commission attributed over 58,000 neonate deaths to drug resistant bacterial infections.[3]

Low quality medicines and their contributions to the burden of AMR are often overlooked, but they are a very real threat. Falsified and substandard medicines can contribute to antimicrobial resistance by being partially or completely ineffective at treating illnesses and infections, and worse, providing sub-therapeutic exposure of medicines to surviving microbes that can drive the development of resistance. This sub therapeutic exposure can also lead to more virulent, deadlier disease.[4]

The impact of poor quality antimicrobials is already being felt around the world, contributing to drug-resistant tuberculosis in Africa, Asia, and other regions. Further, poor quality drugs may undermine the progress in malaria control programs in Latin America, Asia, and other regions[5]. The World Health Organization estimates one in ten medicines in low and middle income countries is substandard or falsified, with antimicrobials representing a significant proportion of the total, thus, contributing to the potential for resistance that could be wide-reaching, both geographically and in multiple classes of medicines.[6]

2. A STRATEGY FOR ANTIMICROBIAL RESISTANCE NATIONAL ACTION PLAN (NAP)

Securing the supply of quality-assured medicines is a fundamental role of any functioning health system, and there are many known solutions, including robust pre-market review, use of public quality standards and targeted post-market surveillance. Similarly, there are many needed steps to global AMR strategies. Securing the quality of medicines therefore, can present one strategy for medicines stewardship under AMR National Action Plans. At the 68th World Health Assembly, member States were urged to develop national action plans to fight AMR by 2017. By one account, 67 Member States, out of 194, have already developed such plans, while 62 others are in the process of doing so.[7]

The government of India's National Action Plan (NAP) on AMR already prioritizes quality of medicine as a key factor in fighting AMR. Published in April 2017,[8] India's NAP details efforts to ensure the quality of antimicrobials available to the public in India using existing solutions (see Table 1). It recommends strengthening the national and state quality control laboratories; regulating unlicensed

Table 1: Summary of the Regulatory Framework for Ensuring the Quality of Antimicrobials in India

Parameter	*Legal Provision*	*Enforcing Authority*	*Description*
Quality Manufacturing	Schedule M of Drugs and Cosmetics Rule, 1945	CDSCO, DoP	• Enforcing appropriate methodology, systems and procedures, manufacturing premises • Ensuring documentation and maintenance for inspection and reference
Quality imports	Drugs and Cosmetics Act Schedule Y (Rules 122A, 122B, 122D, 122DA, 122DAA and 122E) of the Drugs and Cosmetics Rules, 1945	CDSCO, IPC	• Enforcement of standards of quality for imported drugs • Eliminating misbranded, adulterated and spurious drugs
QMS and Shelf life	Schedule M and Schedule P of Drugs and Cosmetic Rules, 1945	CDSCO	• Ensuring documentation and maintenance for inspection and reference • Rules for shelf life of drugs
Pharma-covigilance	Schedule Y, Drugs and Cosmetics Act, 1940	CDSCO, PvPI, MCI	• Drug safety monitoring • Generate independent, evidence based recommendations on the safety of medicines

Source: National Action Plan on Antimicrobial Resistance, Government of India, April 2017.

pharmacies and medical practitioners; establishing quality management systems; strengthening enforcement of regulations and standards; and improving supply chain management of antimicrobials.

Several key regulatory and policy bodies including the Department of Health and Family Welfare, Central Drugs Standards Control Organization (CDSCO), National Aids Control Authority (NACA), Medical Council of India (MCI), Pharmacovigilance Program of India (PvPI), India Pharmacopeia Commission (IPC) and Department of Pharmaceuticals (DoP) have developed short term, medium term and long term plans to ensure the quality of antimicrobials.[8]

In general, the plan emphasizes strengthening manufacturing, monitoring, developing quality assurance systems, and pharmacovigilance.

They fit within the existing WHO framework for combating Substandard and Falsified Medicines. Efforts toward building these capacities to combat AMR will likely have a spill-over effect in improving the systems India need to secure medicines quality more broadly. As countries in the Southeast Asia Regional Office (SEARO) region and beyond look to create and implement their own AMR national action plans, a focus on the quality of antimicrobials as India has done, may present strategies that could help address other medicinal issues as well.

3. CONCLUSION

There is an urgent need to better appreciate and understand the intersections of quality medicines and AMR, as countries make concrete steps towards containing AMR through national action. As part of AMR plans, countries like India are already making plans to build strong, sustainable quality systems to create a better supply of quality-assured medicines, while strengthening post-market quality surveillance to better guarantee that these quality-assured medicines can reach patients. This approach can help build local capacity of manufacturers and regulators and provide the space for solutions to eliminate falsified, substandard and unapproved medicines. Positive end-effects could benefits not only AMR containment efforts, but multiple country health strategies.

REFERENCES

[1] World Health Organization (WHO), Antimicrobial Resistance Global Report on Surveillance (2014). http://apps.who.int/iris/bitstream/10665/112642/1/9789241564748_eng.pdf?ua=1

[2] O'Neill, Jim, Antimicrobial Resistance: Tackling a Crisis for the health and wealth of nations (December 2014); O'Neill, Jim, Safe, Secure, and Controlled: Managing the Supply Chain of antimicrobials (November 2015).

[3] The Lancet Global Health, Antimicrobial resistance—a threat to neonate survival (2016), http://www.thelancet.com/pdfs/journals/langlo/PIIS2214-109X(16)30221-2.pdf

[4] Nwokike, Jude; Clark, Aubrey; Nguyen Phillip P.; "Medicines quality assurance to fight antimicrobial resistance."Bulletin of the World Health Organisation 2018; 96:135-137. doi: http://dx.doi.org/10.2471/BLT.17.199562.

[5] Gaurvika, M.L. Nayyar, Joel G. Breman, Paul N. Newton, James Herrington, Poor-quality antimalarial drugs in southeast Asia and

sub-Saharan Africa, The Lancet Infectious Diseases, Vol. 12, Issue 6, 2012, pp. 488–496, ISSN 1473-3099, https://doi.org/10.1016/S1473-3099(12)70064.

[6] http://www.who.int/news-room/detail/28-11-2017-1-in-10-medical-products-in-developing-countries-is-substandard-or-falsified.

[7] Antimicrobial resistance. Report by the Secretariat. A70/12. Geneva: World Health Organization; 2017. Available from: http://apps.who.int/gb/ebwha/pdf_files/WHA70/A70_12-en.pdf

[8] National Action Plan on Antimicrobial Resistance, Government of India, April 2017. Available from http://www.searo.who.int/india/topics/antimicrobial_resistance/nap_amr.pdf

Phage Lysate Immunotherapy of Brucellosis in Cows

Hari Mohan Saxena* **and Sugandha Raj**
Department of Veterinary Microbiology, College of Veterinary Science, Guru Angad Dev Veterinary and Animal Sciences University (GADVASU), Ludhiana (Punjab), India
E-mail: *drhmsaxena@gmail.com

ABSTRACT: Brucellosis, caused by Brucella organisms, is an important zoonotic disease causing huge economic losses to the livestock industry worldwide. The affected animals remain carrier throughout their lives. Antibiotic resistance has also been reported in Brucella organisms. Currently there is no satisfactory and cost effective treatment for bovine Brucellosis. We have carried out immunotherapy of bovine Brucellosis by phage lysate of Brucella abortus attenuated strains RB51 and S19. A single dose of 2 ml of the S19 lysate or the cocktail lysate, comprising lysates of both RB51 and S19, administered subcutaneously to naturally Brucellosis affected adult cattle induced strong antibody response in cattle. The increase in the titers between 0 and 90 day in case of animals treated with S19 lysate and cocktail (S19 + RB51) lysate was very significant ($P < 0.01$). In case of RB51 lysate immunized animals and cattle which did not receive any treatment, the differences among the titers at various intervals were non-significant.

Keywords: Antibody Titer, Brucella, Brucellosis, Phage Lysate, Standard Tube Agglutination Test, STAT.

1. INTRODUCTION

Brucellosis is endemic in India and is prevalent in all parts of the country. It causes heavy economic loss to the animal industry through delayed conception, late-term abortions and retention of placenta and temporary or permanent infertility (Kollannur *et al.*, 2007) in females and orchitis and epididymitis in males, with excretion of organisms in semen, uterine discharges, and in milk (Godfroid *et al.*, 2013). Numerous outer and inner membrane, cytoplasmic, and periplasmic protein antigens of Brucella have been characterized, yet the antigen that dominates the antibody response is Lipopolysaccharide (LPS). Some antigens are recognized by the immune system during infection and are potentially useful in diagnostic tests. The L7/L12 ribosomal proteins are important in stimulating cell-mediated responses (Oliveira and Splitter 1994). The role of humoral immunity against intracellular bacterial infections is limited and may not be protective. Antibody mediated opsonization by immunoglobulins (IgM, IgG1, IgG2a and IgG3) enhances phagocytic uptake of bacteria, limiting the level of initial infection with *Brucella* but has little effect on intracellular course of infection (Bellaire *et al.,* 2005, Baldwin and Goenka 2006).

We explored the immunotherapeutic potential of phage lysates of S19 and RB51 in Brucellosis in adult cattle and employed STAT to monitor the effect of lysate therapy on the antibody titers in cattle.

2. MATERIALS AND METHODS

2.1 Brucella abortus Vaccine Strains

Live organisms from the attenuated vaccines *Brucella abortus*—strain 19 and strain RB51 (Indian Immunologicals, Hyderabad) were used for making phage lysates for immunotherapy. The bacterial cultures were maintained on *Brucella* agar plates, Trypticase Soy Agar (TSA) and Farrell's medium (OIE 2009) and slants by serial sub-culturing in *Brucella* selective broth on every fortnight and storing the cultures at 4°C.

2.2 Brucella Phage

A broad acting phage lytic to *Brucella* organisms isolated in our laboratory (Gupta and Saxena 2017a) was used for lysing *Brucella abortus* strains S19 and RB51 for making lysates. Phage as a crude, concentrated suspension prepared in SM diluent, was first revived by agar overlay technique as per the procedure reported earlier (Gupta and Saxena 2017b). The phage preparation obtained as mentioned above was amplified to 250 ml master lot using the liquid culture method (Jain *et al.,* 2015).

2.3 Generation of Phage Lysates of Brucella

The methodology for generation of phage lysate preparations against *Brucella abortus* using Brucella phage has been reported earlier

(Jain *et al.*, 2015). The *Total Viable Count (TVC)* of 24–48 hr *Brucella* selective broth culture of the strains S19 and RB51 incubated at 37°C was adjusted to 2 × 10^8 cfu/ml. Opacity of broth was adjusted to Mac Farland's tube no. 3 for the purpose of fetching appropriate antigenic biomass. Phage was added as per optimized *Multiplicity of Infection (MOI)* and TVC of the respective indicator strains (with phage bacteria ratio of 1:50) and the mixture was further incubated for 6–7 hours at 37°C for complete lysis and clearance of turbidity. 100 ml each of phage lysate was prepared against both the strains (*Brucella abortus* S19 and RB51). The phage lysate cocktail was then passed through a 0.1 μm filter (Pall Life Science) to separate out the phage from the lysate and the filtrate was stored in sterilized vials at 4°C. Sterile aluminium hydroxide gel suspension in saline in ratio of 1:10 was added to the preparation. Total protein content of the phage lysate, as determined by Nanodrop spectrophotometer, was 0.58 mg/ml for S19 lysate and 0.64 mg/ml for RB51 lysate. The phage lysate cocktail (combination of *Brucella abortus* S19 and RB51 lysates) was prepared by addition of equal quantity of S19 phage lysate and RB51 lysate. Sterilized 1% Aluminium hydroxide gel suspension in saline was mixed aseptically with the test preparations in ratio of 1:10 (final aluminium concentration 0.1%) and incubated at 37°C for 24 hours and then stored at 4°C. The phage lysates were then subjected to sterility and safety tests.

2.4 Sterility Testing of Lysates

A loopful of the lysate was suspended in 5 ml BHI and BSM broth as well as streaked on BHI and BSM, Trypticase Soy Agar followed by incubation at 37°C. The broth and plates were examined up to 48 hrs for any microbial growth.

2.5 Compliance with Animal Ethics

All the experimental protocols performed on mice and cattle were approved by the *Institutional Animal Ethics Committee (IAEC)* of GADVASU, Ludhiana. Animals were kept in approved facilities as per the CPCSE guidelines under PCA Act. 1960.

2.6 Safety Test of the Phage Lysates in Mice

Safety test of all the preparations of lysates was conducted in mice as recommended in the Indian Pharmacopoeia 2010 before commencement of immunization of cattle. Three groups of three adult healthy mice, each were injected with 0.2 ml volume of lysate (S19 or RB51 or cocktail of both) by subcutaneous route. One group of three adult healthy mice was left untreated as control. The mice were observed for any untoward reaction or mortality up to 7th day of inoculation.

2.7 Immunotherapy Trials in Cattle

Trials of phage lysate therapy were carried out on 21 naturally Brucellosis affected adult cows. The Brucellosis positive adult cows were divided into four groups:

Group I Animals (n = 6) were immunized with *Brucella abortus* S19 lysate only,

Group II Animals (n = 5) were immunized with *Brucella abortus* RB51 lysate only,

Group III Animals (n = 5) were immunized with cocktail of both S19 and RB51 lysates. A dose of 2 ml lysate was administered through subcutaneous route whereas

Group IV (n = 5) served as a control, and received no immunization.

Blood samples from cattle were collected through jugular vein at 0 day and at 30, 60, 75 and 90 days post treatment for studying the immune response of the animals specific to the *Brucella* organisms. Sera were separated from blood and stored at –20°C and –80°C, respectively till further use.

2.8 Rose Bengal Plate Test (RBPT)

Equal volumes (25 µl each) of RBPT colored antigen (Punjab Veterinary Vaccine Institute, Ludhiana) and test serum were mixed on a clean glass slide (Morgan *et al.,* 1978). The slide was observed till 3 min. for formation of clumps. Formation of clumps indicated a positive reaction while the absence of clear clumps was considered as a negative reaction.

2.9 Estimation of Antibody Titers by Standard Tube Agglutination Test (STAT)

The standard method recommended by OIE (2004) was followed. Ten agglutination tubes were labelled and placed in rack. Then 0.8 ml of 0.5% carbol saline was added to the first tube of the series. 0.5 ml of carbol saline was added to the rest of the tubes except 8, 9 and 10 which were kept as controls, containing 1.25, 1.50 and 1.75 ml of carbol saline, respectively. 0.2 ml serum was added to the first tube and the contents were mixed. 0.5 ml of plain *Brucella abortus* antigen (Punjab Veterinary Vaccine Institute, Ludhiana) was added to the first tube and final dilutions were made in ratios of 1:10, 1:20, 1:40, 1:80 and so on. To the control tubes 8, 9 and 10, *Brucella abortus* plain antigen 0.75, 0.50 and 0.25 ml, respectively was added. Tubes were incubated for 24 hours at 37°C and results were compared with the antigen control tube showing 50% agglutination. The highest serum dilution showing 50% agglutination was taken as the end point for the titer serum. A titer of 1:40 or above was considered positive (see Table 1).

Table 1: STAT Protocol Adapted in this Experiment

<table>
<tr><th>Tube No.</th><th>Carbol Saline (ml)</th><th>Test Serum (ml)</th><th>B. abortus Plain Antigen (ml)</th><th>Final Dilution</th></tr>
<tr><td>1</td><td>0.8</td><td>0.2</td><td>0.5</td><td>1:10</td></tr>
<tr><td>2</td><td>0.5</td><td rowspan="9">Serial dilution was performed after thorough mixing. 0.5 ml of the contents was transferred from tube no.1 to the next tube up to tube no. 7. Finally 0.5 ml of the contents was discarded from tube no. 7</td><td>0.5</td><td>1:20</td></tr>
<tr><td>3</td><td>0.5</td><td>0.5</td><td>1:40</td></tr>
<tr><td>4</td><td>0.5</td><td>0.5</td><td>1:80</td></tr>
<tr><td>5</td><td>0.5</td><td>0.5</td><td>1:160</td></tr>
<tr><td>6</td><td>0.5</td><td>0.5</td><td>1:320</td></tr>
<tr><td>7</td><td>0.5</td><td>0.5</td><td>1:640</td></tr>
<tr><td>8</td><td>1.25</td><td>0.75</td><td></td></tr>
<tr><td>9</td><td>1.50</td><td>0.50</td><td></td></tr>
<tr><td>10</td><td>1.75</td><td>0.25</td><td></td></tr>
</table>

Controls: 1. Tube no. 8—25% agglutination, 2. Tube no. 9—50% agglutination, 3. Tube no. 10—75% agglutination.

2.10 Statistical Analysis

Data pertaining to serum antibody titers by STAT was statistically analyzed by t-test and ANOVA.

3. RESULTS AND DISCUSSION

3.1 Immunization of Brucellosis Affected Cattle with Phage Lysate

Brucellosis affected cattle were treated with *Brucella abortus* strain 19 and RB51 phage lysate vaccines. Pre (0 day) and post treatment serum samples were collected at 0, 30, 60, 75 and 90 days and stored at –20°C and –80°C respectively until used. All the serum samples from the infected animals under study showed positive reaction with RBPT during the entire period since the antibody developed against *Brucella* organisms gives the agglutination reaction with the Rose Bengal coloured antigen. Antibody titers showed a classical pattern of gradual rise from prevaccation levels to achievement of peaks followed by plateau and then gradual decline over a period of time.

3.2 Titers by STAT

Titers of antibody induced by the phage lysate vaccines were estimated by STAT. The serum antibody titers by STAT are presented in Tables 2–5, and Figure 1. In case of S19 lysate immunized animals, the mean

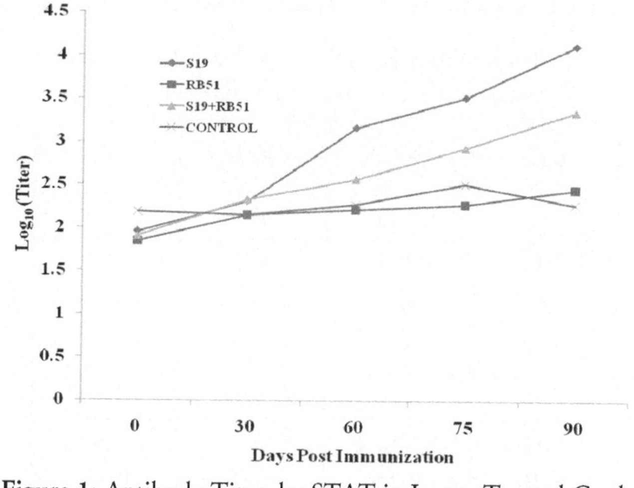

Figure 1: Antibody Titers by STAT in Lysate Treated Cattle

Table 2: Antibody Titers by STAT in Brucellosis Affected Cattle Treated with S19 Lysate

Animal No.	*0 D*	*30 D*	*60 D*	*75 D*	*90 D*
1469	2.204	2.806	3.107	3.408	4.010
1456	1.602	1.903	3.107	3.709	4.311
1405	1.903	2.204	2.806	3.107	3.709
1397	2.505	2.806	3.709	3.709	4.010
1376	1.602	1.903	3.107	3.408	4.311
1440	1.903	2.204	3.107	3.709	4.311
Mean±SD	**1.953±0.35**	**2.304±0.41**	**3.157±0.29**	**3.508±0.24**	**4.110±0.24**

Table 3: Antibody Titers by STAT in Brucellosis Affected Cattle Treated with RB51 Lysate

Animal No.	*0 D*	*30 D*	*60 D*	*75 D*	*90 D*
1479	1.602	1.903	2.204	2.204	2.505
1444	1.602	1.903	1.903	2.204	2.204
1495	2.505	2.806	2.806	2.505	2.505
1281	1.602	1.903	1.903	1.903	2.204
1385	1.903	2.204	2.204	2.505	2.806
Mean ± SD	**1.842 ± 0.39**	**2.143 ± 0.39**	**2.204 ± 0.36**	**2.264 ± 0.25**	**2.444 ± 0.25**

Table 4: Antibody Titers by STAT in Brucellosis Affected Cattle Treated with Cocktail (S19+RB51) Lysates

Animal No.	*0 D*	*30 D*	*60 D*	*75 D*	*90 D*
1367	1.602	1.903	2.505	2.806	3.107
1390	1.903	2.204	2.505	3.107	3.709
1418	2.204	2.806	2.806	3.107	3.408
1460	1.602	1.903	2.204	2.505	2.806
1471	2.204	2.806	2.806	3.107	3.709
Mean ± SD	**1.903 ± 0.30**	**2.324 ± 0.45**	**2.565 ± 0.25**	**2.926 ± 0.26**	**3.347 ± 0.39**

Table 5: Antibody Titers by STAT in Brucellosis Affected Untreated Cattle

Animal No.	*0 D*	*30 D*	*60 D*	*75 D*	*90 D*
1408	1.903	1.602	1.903	1.903	2.204
1493	2.505	2.204	2.204	2.806	2.505
1351	2.806	2.505	2.505	2.806	2.505
1319	2.806	2.806	2.505	2.806	2.204
1382	1.903	1.602	1.903	2.204	1.903
Mean ± SD	**2.183 ± 0.55**	**2.143 ± 0.53**	**2.264 ± 0.39**	**2.505 ± 0.42**	**2.264 ± 0.25**

titers ranged from 1.953 ± 0.35 on 0 day to 2.304 ± 0.41 on 30 day, 3.157 ± 0.29 on 60 day, 3.508 ± 0.24 on 75 day and 4.110 ± 0.24 on 90 day, respectively. The increase in the titers from 75 day to 90 day was significant ($P < 0.05$). The increase in the titers between 0 and 60 day, 0 and 75 day and 0 and 90 day, between 30 and 60 day, 30 and 75 day, 30 and 90 day and between 60 and 90 day was found to be very significant ($P < 0.01$).

In case of RB51 lysate immunized animals, the mean titers ranged from 1.842 ± 0.39 on 0 day to 2.143 ± 0.39 on 30 day 2.204 ± 0.36 on 60 day, 2.264 ± 0.25 on 75 day and 2.444 ± 0.25 on 90 day. The titers at various intervals did not vary significantly. In case of animals treated with cocktail (S19 + RB51) lysate vaccine, the titers ranged from 1.903 ± 0.30 on 0 day to 2.324 ± 0.45 on 30 day, 2.565 ± 0.25 on 60 day, 2.926 ± 0.26 on 75 day, and 3.347 ± 0.39 on 90 day, respectively. It was observed that, the increase in the titers from 0 to 60 day, 30 to 75 day, and between 60 and 90 day was significant ($P < 0.05$). On the other hand, difference in titers from 0 to 75 day, 0 to 90 day and between 30 and 90 day was very significant ($P < 0.01$).

In Brucellosis affected cattle which did not receive any treatment, the titers ranged from 2.183 ± 0.55 on 0 day to 2.384 ± 0.45 on 30 day, 2.264 ± 0.39 on 60 day, 2.505 ± 0.42 on 75 day and 2.264 ± 0.25 on 90 day respectively. The differences among the titers at various intervals were non-significant. The results indicate that the antibody titers enhanced considerably by *Brucella abortus* S19 lysate, similar to the earlier reports with whole live S19 vaccine. Interestingly the cocktail lysate generated a robust antibody response which consistently increased very significantly over the 3 months' period of observation.

4. CONCLUSIONS

Phage lysate of *Brucella abortus* attenuated strain RB51 alone or in combination with lysate of *Brucella abortus* attenuated strain S19 administered sub-cuteneously at a single dose of 2 ml in naturally Brucellosis infected adult cows led to significant increase in antibody titers at various intervals as estimated by STAT.

REFERENCES

[1] Baldwin, C.L. and Goenka, R. (2006). Host immune responses to the intracellular bacterium *Brucella*: does the bacterium instruct the host to facilitate chronic infection? *Critical Reviews in Immunology,* 26: 407–442.

[2] Bellaire, B.H., Roop, R.M. and Cardelli, J.A. (2005). Opsonized virulent *Brucella abortus* replicates within nonacidic, endoplasmic reticulum-negative, LAMP-1-positive phagosomes in human monocytes. *Infection and Immunity* 73: 3702–3713.

[3] Godfroid, J., Garin-Bastuji, B., Saegerman, C. and Blasco, J.M. (2013). Brucellosis in terrestrial wildlife. *Revue Scientifique et Technique del Office International des Epizooties* 32: 27–42.

[4] Gupta, V. and Saxena, H.M. (2017a). A new bacteriophage based luminescence assay for diagnosis of Brucellosis. *Ind J Exp Biol* 55: 296–302.

[5] Gupta, V. and Saxena, H.M. (2017b). Isolation and characterization of BpL1, a broad acting lytic bacteriophage against *Brucella. Int J. Curr Microbiol App Sci.,* 6: 2486–2496.

[6] Jain, L., Rawat, M., Prajapati, A., Tiwari, A.K., Kumar, B., Chaturvedi, V.K., Saxena, H.M., Ramakrishnan, S. and Kumar, J. (2015). Protective immune response of aluminium hydroxide gel adjuvanted phage lysate of *Brucella abortus* S19 in mice against direct virulent challenge with *B. abortus* 544. *Biologicals,* 43: 369–376.

[7] Kollannur, J.D., Rathore, R. and Chauhan, R.S. (2007). Epidemiology and economics of Brucellosis in animals and its zoonotic significance. *Proceedings of XIII International Congress in Animal Hygiene. International Society for Animal Hygiene,* pp. 466–68.

[8] Morgan, W.J., Mackinnon, D.T., Gill, K.P.W., Gower, S.G.M. and Norris, P.I.W. (1978). Brucellosis Diagnosis: Standard Laboratory Techniques Report Series No. 1. MAFF, Weybridge, England.

[9] OIE (2004). Bovine Brucellosis, section 2.3 Manual of standards for diagnostic test and vaccines. 5th Edn. Office International Des Epizooties, Paris, France.

[10] OIE (2009). Bovine Brucellosis Section 2.4.3. Manual of the Diagnostic Tests and Vaccines for Terrestrial animals, Vol. 1, 5th Edn. Office International Des Epizooties, Paris, France.

[11] Oliveira, S.C. and Splitter, G.A. (1994). Subcloning and expression of *Brucella abortus* L7/L12 ribosomal gene and T. lymphocyte recognition of the recombinant protein. *Infection and Immunity,* 62: 5201–5204.

Antimicrobial Activity of Actinomycetes of Soils of Mongolia

J. Norovsuren

Laboratory of Microbiology, The Institute of General and Experimental Biology, Mongolian Academy of Sciences, Ulaanbaatar, Mongolia

E-mail: norvo@mail.ru

ABSTRACT: The paper shows the result of the Antimicrobial Activity of Actinomycetes of Soils of Mongolia. Soil is the habitat of many types of microorganisms. Through the soil, pathogens of tuberculosis, plague, dysentery, typhoid fever, cholera, gas gangrene, botulism, anthrax and others can be transmitted. It has been experimentally proved that a number of pathogenic microorganisms can survive for a long time or even multiply in the soil. In this work, population and taxonomy of actinomycetes in Mongolian soils were studied. The purpose of the research work was to search for actinomycetes synthesizing biologically active substances with antagonistic activity against microbes, phyto-pathogenic fungi and bacteria that cause dangerous diseases of agri-cultural plants.

Keywords: Rare Genera Actinomycetes, Mongolia, Antagonism, Economic Valuable Microorganisms.

1. INTRODUCTION

At present, representatives of almost all known genera of action-mycetes are isolated from the soil or found in it. Among the actinomycetes, many forms have been identified that are produced by antibiotics used in plant growing, animal husbandry and in some fermentation plants.

The actinomycetales species produces over 10000 bioactive com-pounds, 75% derived from Streptomyces. By the variety and the amount of antibiotics formed, actinomycetes occupy the first place among microorganisms.[1]

Soil is the habitat of many types of microorganisms. Through the soil, pathogens of tuberculosis, plague, dysentery, typhoid fever, cholera, gas gangrene, botulism, anthrax and others can be transmitted. It has been experimentally proved that a number of pathogenic micro-organisms can survive for a long time or even multiply in the soil.

Economic valuable microorganisms can serve as the basis for the creation of cost-effective production of a means of protecting plants and agricultural products, medicines for animals and humans. Consequently, the search and investigation of new properties of metabolites of microbiological origin is an actual task of modern biology.

The study of antagonism represents an extensive field of research, both for clarifying its biological essence, and for the purpose of practical use in medicine and veterinary medicine.

The purpose of our research work is to search for actinomycetes synthesizing biologically active substances with antagonistic activity against microbes, phytopathogenic fungi and bacteria that causes dangerous diseases of agricultural plants.

2. MATERIALS AND METHODS

Objects of the study were samples of soils of mountain-forest, mountain-steppe, steppe, desert and meadow ecosystems of Mongolia. Samples were taken from the upper horizon of virgin soils (5–20 cm). Isolation and differentiated enumeration of actionmycetes were carried out by plating the dilutions of soil suspensions on the solid nutrient media (humus-vitamin agar and the medium with sodium propionate).[13] For inoculations, air dry weighted soil samples were used. The media were supplemented with nystatin in order to suppress fungal growth and with the vitamin B complex (4 μg/l) to improve actinomycete recovery.

The samples were incubated for 3–4 weeks at 28°C; whereafter, the numbers of actinomycetal colonies were counted. They were subdivided into colony-forming units (CFU/g) and the relative abundance of the species were determined in each sample. The Gauze 2 medium was used for the isolation of pure actinomycetal cultures and their further cultivation.[2] The preliminary identification of the isolated strains was performed using the Bergey manual according to the following morphological characteristics: the fragmentation and branching of mycelium, the presence and location of spores on air and/or substrate mycelium and their number in chains, the diameter and mobility of spores, the presence of sporangia, and the chemotaxonomic features (the presence of LL- or mesodiaminopimelic acid[11,12] and differentiating sugars[4,5] in hydrolyzates of whole cells). The belonging of *Streptomyces* species to particular sections was identified using the manual by Gauze *et al.*[2]

The antagonistic activity was determined by the diffusion method in agar, the diameter (mm) of the zone of no growth to the test organisms was measured.

3. RESULTS AND DISCUSSION

Extreme environmental conditions are expressed in a sharply continental climate with high daily and annual temperature amplitudes, frequent hurricane winds with high solar radiation; it also manifests itself in the structure and spatial distribution of geosystems, being an important factor in the formation of natural complexes and their unique properties. The soils of Mongolia have their own distinctive features, conditioned by the unique nature of the country.

The number of actinomycetes in all the soil samples investigated reaches tens of thousands of CFU/g of soil depending on the soil type.

The selective method of isolating actinomycetes from soil with a preliminary heating of soil samples (120°C, 1 h) turned out to be efficient for the selective isolation of species of rare genera.

In the soils of Mongolia, representatives of the genera of actinomycetes *Streptomyces, Micromonospora, Saccharopolyspora, Saccharomonospora, Actinomadura, Streptosporangium, Dactylosporangium* and *Nocardia* are found, with only streptomycetes and micromonospores dominating, the remaining genera are rarely found. The study of the taxonomic composition of actinomycetes of the main soil types in Mongolia makes it possible to conclude that the complexes of soil actinomycetes reflect the peculiarities of the natural conditions and specific features of the country's soils and can serve as an indicator of the state and nature of Mongolia's ecosystems.[3, 6–9]

The more antagonists in the soil, the less there are corresponding pathogens of infections. All that contributes to the development of actinomycetes-antagonists, speeds up the healing of the soil.

Antimicrobial substances are a powerful weapon for actinomycetes-antagonists, which they develop to fight their competitors. Each type of actinomycetes-antagonists has its own antimicrobial specificity, i.e. ability to suppress a set of certain types of microorganisms. The ability of actinomycetes to act selectively strictly is constant; it is conditioned by natural or species properties and as such can serve as one of the essential features of the species.

As a result of studying the antimicrobial and antipathogenic properties of soil isolates of actinomycetes isolated from various soil ecosystems in Mongolia, the following test organisms suppress: *Bacillus subtilis (6.0–15 mm), Escherichia coli (7.0–18 mm), Staphylococus aureus (7,0–28 mm), Saccharomyces cerevisiae (7.0–25 mm), Aspergillus (niger) awamori (7.0–22 mm), phytopathogenic fungi and bacteria: Alternaria raphanii brassicae (12.0–24.0 mm), Cladosporium cucumerinum Ell.et.Arth (10–12 mm) and Fusarium oxysporum f.sp. lycopersici (10.0–26.0 mm).*

Mongolia is a developing country for which agricultural production is very important. It is known that plant diseases are widespread and cause significant harm to the agriculture. To combat them is to use a chemical method. Chemical means of protection is, first of all, poisonous substances. In the following years, the negative aspects of its application became visible and real.

Antagonists are the most promising and reliable helpers of a person when using the biological method of struggle. Biopreparations are obtained as a result of long-term scientific research of microorganisms and their influence on the environment, careful study of natural mechanisms and the principle of ecosystem functioning, developmental peculiarities and plant metabolism. It is absolutely environmentally friendly and enhances the ecosystem itself.

In the literature, it has been repeatedly reported that secondary metabolites of actinomycetes phytopathogenic antagonism are the basis for new biologics.[6, 10]

Thus, as a result of screening of actinomycetes from Mongolia's soils with antagonistic activity against pathogens of dangerous diseases of agricultural plants, we have isolated strains of actinomycetes that are promising for further study and for creation of new biopreparations on its basis.

Soil antagonists regulate the composition of soil microflora, play an important role in its improvement ("self-cleaning").

4. CONCLUSION

Population and taxonomy of actinomycetes in Mongolian soils were studied. Actinomycetes isolated from Mongolian soils with antagonistic activity against pathogens of dangerous diseases of agricultural plants can be potential producers of antimicrobial metabolites for the production of new ecologically safe biologics and for the development of new, practically valuable antibiotic drugs in medical practice.

ACKNOWLEDGEMENTS

This work was carried out within the framework of the thematic plan of the RAS and the MAS "Creation of a collection of biotechnologically promising cultures of actinobacteria isolated from Mongolian soils" at the Winogradsky Institute of Microbiology, Russian Academy of Sciences, Moscow, Russia and All-Russian institute of plant protection (FSBSI VIZR) St. Petersburg – Pushkin, Russia.

REFERENCES

[1] Berdy, J. (2005). Bioactive Microbial Metabolites. J. Antibiot. 58 (1), pp. 1–26.

[2] Gause, G.F., Preobrazhenskaya T.P., Sveshnikova M.A., Terekhova L.P. and Maksimov T.C. (1983). Determinant of actinomycetes (in Russian). Moscow. Science, 245 P.

[3] Kurapova, A.I., Zenova, G.M., Sudnitsyna, I.I., Kizilovab, A.K., Manucharova, N.A, Norovsuren, J. and Zvyagintsev, D.G. (2012). Thermotolerant and Thermophilic Actinomycetes from Soils of Mongolia Desert Steppe Zone. Mikrobiologiya. Vol. 81, No. 1, pp. 105–116.

[4] Hasegawa, T., Takizawa, M. and Takida, S. (1983). A rapid analysis for chemical grouping of aerobic actinomycetes// J. Gen. Appl. Microbiol., 1983, V. 29. P. 319-322

[5] Lechevalier, M.P. and Lechevalier, H.A. (1970). Composition of whole – cell hydrolysates as a criterion in the classification of acrobic actinomyces// The Actinomycetales. Eds. H. Prauser, Jena: Gustav Fesher Verlag. pp. 311–316.

[6] Norovsuren, J. (2001). Soil actinomycetes of rare genera from dominant ecosystems of Mongolia. Russian State Agrarian University – MTAA. The thesis for a doctor's degree (Ph.D., Biology). Moscow, 144 P.

[7] Norovsuren, J. (2007). Ecology rare actinomycetes in soils of Mongolia and their role for soil formation. Russian State Agrarian University – MTAA. The thesis for a doctor's degree (Sc.D., Biology). Moscow, 419 P.

[8] Norovsuren, J. (2009). Geographical distribution actinomycetes in soils of Mongolia. (in Russian) Moscow, Russia. Russian State Agrarian University – MTAA, 170 P.

[9] Nororsuren, J., Kuznetsova, A.I., Ecophysiology of actinomycetes in different types of soils in Mongolia. International Conference To the 120th anniversary of the establishment of the Department of Microbiology and to the 150th anniversary of Prof. N.N. Khudyakova "Modern Aspects of Agricultural Microbiology". Moscow, 2016, pp. 49–50.

[10] Norovsuren, J., Opportunity of using properties of actinomycetes in soil biotechnology, presented at conference on Development of Biotechnology and Use of Genetic resources, Ulaanbaatar, Mongolia, 2018, pp. 199–202.

[11] Schaal, K.P. (1987). "Identification of clinically significant actinomycetes and related bacteria using chemical techniques". Germany.

[12] Staneck, J.L. and Roberts, G.P. (1974). Simplified approach to identification of aerobic actinomycetes by thin-layer chromotography // Appl. Microbiol, Vol. 23, pp. 226–231.

[13] Zenova, G.M. (2000). Soil Actinomycetes of Rare Genera (in Russian). Moscow. Lomonosov Moscow State University, p. 81.

Natural Preservatives in Food System to Combat Antimicrobial Resistance

Manish Kumar Chatli*, Amanpreet Kaur, Nitin Mehta, P. Kumar, O.P. Malav and Rajesh V. Wagh

Department of Livestock Products Technology, College of Veterinary Science, Guru Angad Dev Veterinary and Animal Sciences University (GADVASU) Ludhiana (Punjab), India

E-mail: *manishchatlilpt@gmail.com

ABSTRACT: Antimicrobials derived from plant, animal or microbial sources may significantly improve human and animal health. However, the indiscriminate use of antibiotics results in the development of antimicrobial resistance in bacteria. Food may act as a vector for the transfer of antimicrobial resistant bacteria and antimicrobial resistance genes to humans. A strong link between the use of antibiotics during agricultural production and antimicrobial resistance of human pathogens has already been established in number of cohort studies. Food is considered as one of the possible transfer routes. For example, pork and poultry meat can both be sources of transfer of antimicrobial resistant Salmonella typhimurium strains to humans. A study estimated the probability of exposure to 1,000 colony forming units of cephalosporin-resistant Escherichia Coli (CREC) through consumption of a meal containing chicken meat. Fisheries and aquaculture is also playing an important role in spreading antimicrobial resistance in human pathogens.

Keywords: Minimal Inhibitory Concentration, Cephalosporin Resistant Escherichia Coli.

1. ANTIMICROBIAL RESISTANCE: DEFINITION AND MECHANISM INVOLVED

Antimicrobial resistance is the capacity of a microorganism to resist the growth inhibitory or killing activity of an antimicrobial beyond the normal susceptibility of the specific bacterial species. Antimicrobials include antibiotics used for treating infections as well as chemical biocides, which are used for the disinfection of the food processing environment. A microorganism can acquire resistance to an antimicrobial to which it was previously sensitive, meaning that the antimicrobial will no longer be able to kill or inhibit the growth of the microorganism at the same level as before. In general, three types of resistance have been described. Microbiological resistance (*in vitro*) means a reduced susceptibility of bacteria to antibiotics above a breakpoint that is defined by the upper limit of normal susceptibility of the concerned species, which is also called epidemiological resistance. The microbiological resistance can often be confirmed genotypically by demonstrating the presence of a certain antimicrobial resistance gene or resistance mechanism via molecular techniques. Secondly, there is the pharmacological resistance which is based on pharmacokinetic parameters and the normal susceptibility of a bacterial species. If the minimal inhibitory concentration (MIC) of the antibiotic for the bacteria concerned is within the concentration range that can be attained by that antimicrobial, it is susceptible. If the MIC of the antibiotic for the concerned bacteria is higher than the concentration that can be attained at the site of infection, then the bacterium is regarded as resistant. Finally, clinical resistance (*in vivo*) means an infection with the concerned bacterium cannot be treated appropriately anymore and treatment failures are evident (De Roever, 1998).

The underlying mechanisms for the development and propagation of antimicrobial resistance in bacteria is quite complex. It can be by enzymatic degradation of antibiotics, antibiotic target modification, changing the bacterial cell wall permeability and alternative pathways to escape the activity. Enzymatic degradation or modification of antibiotics is a very common mechanism of resistance. Examples are the β-lactamase enzymes hydrolyzing the β-lactam ring of β-lactam antibiotics such as the cephalosporins, which are mainly of concern in Gram-negative bacteria. Resistance by target modification implies a modification of the target molecule of the antibiotic, in general an enzyme, so that the antibiotic loses its binding capacity and hence, its activity. Changing the cell wall or cell envelope permeability implies reducing entry or increasing the efflux of antibiotics, thereby regulating the internal concentration of antibiotics in the cell. Changes in pores can alter or inhibit the entering capability of antibiotics into the cell. Cells can become resistant by deviating from their normal physiological pathway by including an alternative step e.g. by an extra enzyme (Cosgrove *et al.*, 2003; Smith and Coast, 2013).

2. ANTIMICROBIAL RESISTANCE IN FOOD SYSTEMS: AN ESCALATING PROBLEM

Food may be contaminated with antimicrobial resistant bacteria and/or antimicrobial resistance genes in many ways. Antimicrobial resistant bacteria may be found in the soil, in the water and in human or animal fecal material. Animal products may contain antimicrobial resistant bacteria as a result of fecal contamination during slaughter. Plant products may be contaminated with antimicrobial resistant bacteria during production following the use of irrigation water contaminated with human and/or animal feces or by sewage discharges. Food may also be contaminated by the environment, generally referred to as post-contamination. Finally, food can be cross contaminated with antimicrobial resistant bacteria and/or antimicrobial resistance genes originating from other foods during the handling of the food by the consumer (Founou *et al.*, 2016).

The alarming reports of presence of antibiotic residue in livestock products including meat and eggs due to consistent and irrational use of sub-therapeutic dose of antibiotic in the feed of food animals as growth promoters has created a lot of concern among the consumers and researchers. Over the past 30 years, there has been a serious debate regarding the use of antimicrobials in food animals and the subsequent health implications. The concern with the use of antibiotics in slaughter animal is not much attributed to the use in veterinary medicine, but mainly due to the continuous use of antibiotics at sub-therapeutic level for growth promotion and disease prevention (Chang *et al.*, 2014). Foods especially the animal products may be an important vector in the transfer of antibiotic resistant food-borne pathogens which may gain entry at the farm level, or presence of resistance genes in cultures that are intentionally added during processing or due to cross contamination during processing (Verraes *et al.*, 2013). Under certain preservation and processing conditions, sub-lethally damaged or stressed cells may be maintained in the food and pose public health risk.

As per estimates (2015–16) USA alone is consuming 13.5 million kg antibiotics in animal food production and preservations, 0.15 million kg in aquaculture and 0.15 million kg in pet foods. This is almost 400 times the antibiotics used in total human interventions (3.29 million kg). In India, Penicillin is the most commonly used antibiotics in animal foods (731 tons/annum) and closely followed by tetracycline

(685 tons/annum). Presently, the global antibiotic market is valued as US$ 41.2 billion. If no efforts have been made the global (Watkins and Bonomo, 2016) consumption of antibiotics will increase to 127 billion dose/day by 2030.

Further, during production process of certain food products, micro-organisms which can contain antimicrobial resistance genes are intentionally added for obvious reasons e.g. starter cultures, probiotics, biopreserving microorganisms, bacteriophages etc. Starter cultures are added to foodstuffs in order to induce the onset of fermentation and Lactic Acid Bacteria (LAB) are mostly used for that purpose. Some starter cultures have probiotic features too and used in development of yogurt, probiotic sausages etc. They can also be used in biopreservation, however, on contrary, antimicrobial resistance is reported in fermented foods and probiotic strains too. Among the LAB isolated from spontaneously fermented foodstuffs, resistance is most common in *Enterococcus.* In most cases, this bacterium is resistant to vancomycin although resistances to tetracycline, erythromycin and chloramphenicol have also been observed. The prevalence of antimicrobial resistance genes in bacteriophages is largely unknown. Colomer- Lluch *et al.* (2011) reported three genes conferring resistance to β-lactam-antibiotics, *i.e.*, two β-lactamase-genes (*bla*TEM and *bla*CTX-M9) and one gene coding for a modified penicillin-binding protein (*mecA*) in all 30 bacteriophage DNA samples from urban waste water and river water. They concluded that bacteriophages could be environmental reservoirs of antimicrobial resistance genes and can lead to horizontal transfer.

3. NATURAL PRESERVATIVES AS AN ALTERNATIVE TO COUNTER ANTIMICROBIAL RESISTANCE

The World Health Organization reports that, annually, unsafe food results in the illnesses of at least 2 billion people worldwide and can be deadly. Thermal processing is a common method of destroying vegetative microorganisms to ensure food safety, but this technique may cause undesirable nutritional and quality effects. Preservatives are commonly used to reduce the risk of food-borne illnesses but increasing incidences of resistance of major food-borne pathogens and increasing regulatory restrictions and consumer negative response to use of antibiotics in foods have contributed to the pressure for the development of alternative compounds for use as antimicrobial agents. The consumers are looking for foods that are minimally processed, microbiologically safe and chemical free with longer shelf-life. In order to meet the consumers' demands, the food researchers and the food industry as a whole are looking for natural alternatives that exhibit strong antioxidant and antimicrobial activity (Aziz and Karboune, 2018). The use of natural preservatives is not new in the history of food science. The natural food preservatives can be classified into antimicrobial agents and antioxidant agents (Sharif *et al.*, 2017). Alternatives to traditional antimicrobial compounds include bacteriophages, antimicrobial phytochemicals, and antimicrobial peptides. There is less extensive research on the application of antimicrobial compounds from sources other than bacteria.

4. ANTIMICROBIAL AGENT—MICROBIAL ORIGIN

4.1 Bacteriophages

Bacteriophages natural predators of bacteria are ubiquitous in the environment (soil and water) and many food products. These are predominantly present in oceans and counts range from 10^7 to 10^8 phage particles per millilitre in coastal seas and non-polluted water. The estimated phage population in the world is around 10^{31} particles, which makes them the most abundant living objects on our planet. Bacteriophages are specific to one or a limited number of bacterial groups. Normally they are named after the bacterial group, strain, or species that they can kill. For example, the phages that infect the bacterium *Escherichia coli* are called coliphages. Thus, developing pathogen specific antimicrobial substances may improve the antimicrobial efficacy by reducing interactions with non-targeted microbes. Further, non-pathogenic microbes are necessary in the production of some dairy and fermented foods. Additionally, the commensal bacteria present may have benefits for human health such is the case with probiotics microbes or even controlling the growth of pathogenic bacteria.

There are various methods of application of phages along the food chain (Greer, 2005):

- Phage therapy includes reduction of pathogen colonization and is a strategy followed before slaughter or during animal growth to reduce the probability of cross-contamination during food processing.

- Biocontrol involves direct application of phages on the food surfaces.
- Biosanitation involves use of phages in destruction of biofilms formed on the surface of equipment by pathogenic bacteria.
- Phages serve as biopreservation agents since they are able to lyse hosts at temperatures as low as 1°C limiting growth of pathogenic and spoilage bacteria on refrigerated foods and phages can further control their proliferation when taken to room temperature.

Phages have advantages as biocontrol agents for several reasons (Endersen *et al.*, 2014; Verheust *et al.*, 2010; Goodridge and Bisha *et al.*, 2011; Kropinski *et al.*, 2012; Teng-Hern *et al.*, 2014)):

(a) Phages have high specificity to target host cells attributed to bacterial cell wall receptors, thus did not affect the remaining microbiota. This property favors phages over other antimicrobials that can cause collateral damage of commensal microbiota.
(b) Phages have a property of self-replication and self-limiting, meaning that low or single dosages will multiply as long as there is still a host threshold present.
(c) As bacteria develop phage defense mechanisms for their survival, phages continuously adapt to these altered host systems.
(d) Low inherent toxicity, since they consist mostly of nucleic acids and proteins.
(e) Phages are relatively cheap and easy to isolate and propagate.
(f) They can generally withstand food processing environmental stresses (including food physiochemical conditions).
(g) They have prolonged shelf life.
(h) Availability of phages is easier and are readily available in abundance in various foods and have been isolated from a wide variety of raw products (e.g., beef, chicken) processed food, fermented products (e.g., cheese, yoghurt), and seafood (e.g., mussels and oysters).

5. GUIDELINES RELATED TO USE OF PHAGES IN FOOD INDUSTRY

1. Proper Selection of Bacteriophage. In general lytic phages are used for food preservation, whereas lysogenic phage may contain toxic genes which may affect their host.

2. Immunity-Bacteria are known to develop resistance to bacteriophage over time. Therefore, there should be replenishment of new phages with old one.
3. *Industrial Preparation:* The specific nature of bacteriophage infection requires presence of pathogenic host bacteria, if industrial quantities of the phage are to be produced.

6. APPLICATIONS IN FOOD INDUSTRY

E.coli Bacteriophages: *E. coli O157:H7* is a highly virulent food borne pathogen naturally found in the gastrointestinal tract of ruminants. These bacteria are difficult to eradicate and frequently enter the human food chain via contact with contaminated faecal material of animals. Bacteriophage therapy can be used for reducing the presence of these dangerous microbes. The commercial preparation EcoShieldTM manufactured by Intralytix is a blend of three individual phages that provides broad protection specifically against pathogenic strains of E. coli O157:H7.

L. monocytogenes Bacteriophages: Listeria is capable of infecting refrigerated RTE foods. FDA amended the food additive regulations to permit the safe use of a bacteriophage preparation as an anti-listerial agent in RTE meat and poultry products. The preparation (0.1 ppm concentration) consists of a combination of six individual lytic phages, having activity against different *L. monocytogenes* strains. This cocktail is sprayed directly on the surface of the RTE. The bacteriophages on encountering their specific target, L. monocytogenes will trigger a full infection and destruction cycle.

7. COMMERCIAL PHAGE PREPARATIONS

- "ListShield" Intralyx—Effective against different strains of "L. monocytogenes".
- SalmoFresh and SalmoLyse for controlling S. enterica.
- BacWash product line against Salmonella Omnilytics.
- Listex_P100 by Micreos in The Netherlands has been granted Generally Regarded As Safe, 'GRAS' status by the FDA.

8. USE OF COMPETITIVE CULTURES FOR FOOD PRESERVATION

Competitive cultures are the microorganisms which inhibit other pathogenic or spoilage causing microorganisms or reduce their growth in medium onus to their metabolites, indirect (by changing pH, osmotic pressure and surface tension) or direct (by producing toxic component, antimicrobial component, bacteriocin, antibiotic etc.). These cultures are also known as protective or antagonistic cultures. Fermentation technique with known starter culture has emerged as a widely accepted method in this area and lactic acid fermentation is extensively employed for preservation of milk, meat, fruits, vegetables and other raw foods. Fermented food products are known to exist since antiquity though the mechanisms were identified till the turn of last century. Today various strains of lactic acid bacteria are used as starter cultures for fermentation of food items whose shelf-life is enhanced. Growth of food spoilage and pathogenic bacteria is arrested by this technique in two ways. Firstly, there is a competition between fermenting and spoilage or pathogenic groups for nutrients required for growth. Secondly, inhibitors like lactic acid, diacetyl, hydrogen peroxides, microgard, reuterin and bacteriocins elaborated by starter organisms check the growth of these unwanted bacteria. Protective cultures (PCs) performance is affected by temperature, inoculum size, food. Temperature affects growth rate, and bacteriocin production of the microorganisms. Inoculum size affects the production rate and speed of the antimicrobial substances and the sensory quality of the product. Food composition can support or inhibit the protective cultures like salts, spices, added preservatives have a synergetic effect with PCs. Lactic Acid Bacteria (LAB) that comprise of *Lactobacillus, Lactococcus, Leuconostoc* inhibit the growth of pathogenic and spoilage microorganisms, maintain the nutritive quality and improve the shelf-life of food due to the production of inhibitory agents.

9. FOOD PRESERVATION BY BACTERIAL METABOLITES

Bacterial metabolites are the biomolecules produced by various micro-organisms. These include organic acids, enzymes, peptides, bacteriocins etc.

Organic acids viz. lactic acid, butyric acid, propionic acid, formic acid etc. are produced by various bacteria like lactic acid bacteria, *Aceto-*

bacter aceti, Propionibacterium spp. The type and quantity of organic acid varies with the type of microorganisms. Homofermenter LAB produces lactic acid from hexose, whereas heterofermenter produces mixture of acids including butyric acid, propionic acid, formic acid, acetic acid etc. along with gases. Various researches postulated that both lactic acid and acetic acid have wide spectrum antimicrobial effects on fungus, yeast and bacteria and acetic acid has synergistic effect against fungi in the presence of propionic acid.

Hydrogen peroxide (H_2O_2) is produced during aerobic growth of lactic acid bacteria and it is an oxidizing component. Hydrogen peroxide has antimicrobial effect on vegetative cells and spores of most microorganisms. Antimicrobial effect of H_2O_2 has occurred by changing chemical structure and biochemical characteristics of enzymes of target microorganisms. H_2O_2 inhibit most pathogen bacteria like *Staphylococcus, Escherichia coli* O157:H7, *Pseudomonas* and *Clostridium*, also spoilage. It has been used as protective component in region where hygienic conditions are not good and during transport of raw milk.

Diacetyl is a product synthesized from pyruvate as a result of metabolism of hexoses by lactic acid bacteria. Diacetyl has butter odour. It was found in most fermented product, especially dairy products and products from lactic acid fermentation. It shows more inhibitor effect on Gram (–) bacteria, fungus and yeasts than Gram (+) bacteria.

Bacteriocins are the ribosomally synthesized peptides i.e. proteinaceous antimicrobial substances that inhibit the growth of spoilage and pathogenic organisms without toxic or other adverse effect. A recent definition of bacteriocins produced by lactic acid bacteria suggests that they should be regarded as extracellularly released primary or modified products of bacterial ribosomal synthesis. The broad classification of the bactetriocins produced by LAB is as follows:

- **Class 1 Lantibiotics**–which are characterized by some unusual amino acid such as lanthione or beta-methyl lanthionine.
- **Class 2 Small heat stable peptides**–which are small hydrophobic membrane active peptides.
- **Class 3 Large heat labile proteins**–which are of a higher molecular mass.
- **Class 4 Complex proteins**–whose activity requires the associations of carbohydrates or lipid moiety.

Among the various classes of bacteriocin, classes 1 and 2 have extensive role in food preservation. Class 1 bacteriocin that includes nisin, have a dual mechanism in killing their target cells through (a) inhibition of cell wall synthesis and (b) pore formation. Class 2 bacteriocin include-ing Pediocin acts on the cytoplasmic membrane, thereby dissipating ion gradients and inhibiting transport of amino acids in sensitive cells.

10. ANTIMICROBIAL AGENT—PLANT ORIGIN

Herbs and spices have received substantial attention as a phytogenic/phytobiotic alternative to antibiotic growth promoters because of their antibacterial, antioxidant, anti-inflammatory potential, digestive stimulant, immune stimulants, hypocholestrolemic, growth promoting properties (Eevuri and Putturu, 2013). Their use can ultimately minimize/exclude the serious threat of development of bacterial resistance due to use of antibiotic growth promoters (Peric *et al.*, 2009).

Phytochemicals are nonnutritive plant components that confer organoleptic properties and serve as antimicrobial agents. The concentration, composition, structure, and functional groups serve an important role in determining antimicrobial activity. Phenolic compounds are generally the most effective. The most important phytochemicals used as food preservatives are essential oils, which have been used by humans since ages.

Flavonols are phenolic compounds with a carbonyl and a 3-hydroxyl group with a C3–C6 aromatic ring linkage. They are effective against many microorganisms because of their ability to bind to and inactivate proteins and to complex with bacterial cell walls. Quinones are aromatic rings with two carbonyls, providing a stable source of free radicals. In addition to serving as antioxidants, they are potent antimicrobial compounds. Similar to flavones, flavonols, and flavonoids, the antimicrobial mode of action is to bind to and inactivate proteins. In addition, they may make substrates unavailable to the microorganism.

11. ESSENTIAL OILS

Among the natural antimicrobials, essential oils are widely studied as they are known to have antimicrobial, antioxidant, antimutagenic and anticarcinogenic properties. Essential oils, or terpenes, are secondary metabolites and are based on an isoprene structure. They are volatile compounds that provide the fragrance of plants and are mainly responsible for the flavor and aroma of spices. Terpenoids contain additional elements, such as oxygen, and confer activity against bacteria, fungi, protozoa, and viruses. They are volatile aromatic oily liquid usually with a molecular weight less than 300 and obtained from plant material (buds, flowers, leaves, bark, twigs, seeds, herbs, wood, fruits and roots). The essential oils have a number of different chemical compounds present in them that determine their mechanisms for antibacterial activity, but in general it may be due to their hydrophobicity. It has been reported that Eos containing high percentage of phenolic compounds e.g. carvacrol, thymol and eugenol have more antibacterial properties (Ebrahimi and Khosravi-Darani, 2013). As essential oils contain highly diverse groups of phytochemicals, their antimicrobial modes of action have been suggested to involve multiple

Table 1: Antimicrobials of Plant Origin and their Effect on the Food System

Source	*Active Principle*	*Scientific Name*	*Foods/Medium*	*References*
Cinnamon	Cinnamic aldehyde and eugenol	*Cinnamomum zeylanicum*	Inhibitory effect on microbial growth in ground beef	Shaltout *et al*, 2017
Cloves	Eugenol	*Syzygium aromaticum*	Potent antimicrobial in raw chicken meat emulsion	Singh *et al.*, 2014
Garlic	Allicin	*Allium sativum*	Antimicrobial effects on Gram-negative and Gram-positive bacteria and multi-drug resistant strains of *Escherichia coli*	Erginkaya and Konuray., 2017
Blueberry	Anthocyanins	*Vaccinium spp.*	Antimicrobial effect against *Listeria monocytogenes* and *Salmonella* Enteritidis	Shen *et al.*, 2014
Mustard	Isothiocyanate	*Brassica spp.*	Antimicrobial effect on *E. coli* O 157:H7, *Listeria monocytogenes* and *Salmonella enteric* serovar Typhimurium strains.	Kang *et al.*, 2003

targets. The unique hydrophobicity features of essential oils render their abilities to react with lipids on the bacterial cell membranes, increasing the membrane permeability and disturbing the original cell structure.

As the essential oils are highly volatile in nature, nowadays, micro-encapsulation methods are employed to protect them from volatile-zation. Different coating materials are used that entraps the essential oil as a core material, which not only help in protection from volatilization but also allows controlled release of the active ingredients in the food system.

12. ANTIMICROBIAL AGENT—ANIMAL ORIGIN

12.1 Bioactive Peptides

Bioactive peptides are short chain peptides consisting of about 2–30 amino acids in length obtained from their precursor proteins. They are inactive within the parent protein chain and can be released by chemical/enzymatic hydrolysis or by microbial fermentation. Enzymatic hydrolysis is the method of choice for peptide generation by food and pharmaceutical industries as the end peptides lack residual organic compounds and toxic residues (Lafarga and Hayes, 2015). Bioactive peptides can be generated from a wide range of protein rich foods such as eggs, red and white meat, milk, meat by-products, and vegetable proteins like soyabean etc. The enzymatic hydrolysis of bioactive peptides involves the use of proteolytic enzymes derived from animal, microorganisms and plants.

Bioactive peptides are multifunctional in nature exhibiting hormone or drug like activity, which modulate physiological function through binding interactions to specific receptors on target cells leading to induction of physiological responses. The activity of bioactive peptides depends on their amino acid composition as well as the sequence within the parent protein chain. Bioactive peptides exhibit various functional activities viz. antimicrobial, antithrombotic, antihypertensive, opioid, immunomodulatory, mineral binding and antioxidative.

Many of the egg proteins are antimicrobial in nature protecting the yolk from invasion by microorganisms. Antimicrobial peptides are capable of killing a wide range of cells and microbes including bacteria, fungi, protozoa, enveloped virus, malignant cells and parasites

(Thammasirirak *et al.*, 2010). Peptides corresponding to amino acid residues 98–112, 98–108, and 15–21 derived from lysozyme is reported to have antimicrobial activity against *E. coli* and *S. aureus* (Pellegrini *et al.*, 2000; Mine *et al.*, 2004). Ibrahim *et al.* (2000) also reported the antimicrobial activity of OPAT-92 derived from ovotransferrin against Gram negative bacteria. Bovine blood haemoglobin had been used as a source to generate peptides that possess opioid and antimicrobial activities (Adje *et al.*, 2011).

12.2 Enzymes

Lysozyme, derived from eggs, is a single peptide protein active against beta 1–4 glycosidic linkages between N-acetylmuramic acid and N-acetylglucosamine found in peptidoglycan cell wall of both gram-positive and gram-negative bacteria. Unalan *et al.* (2011) reported that zein films incorporated with lysozyme and disodium-ethylene-diamine-tetra-acetic acid (Na_2EDTA) showed antimicrobial activity against *L. monocytogenes, E. coli* O157:H7 and *Salmonella typhimurium* in refrigerated ground beef patties. Lactoperoxidase is a natural enzyme of milk and has three components: lactoperoxidase, thiocyanate, and hydrogen peroxide. The hydrogen peroxide serves as a substrate for lactoperoxidase in oxidizing thiocyanate and iodide ions that result in the generation of highly reactive oxidizing agents. Another enzyme, Lactoferrin, an iron binding glycoprotein also inhibits microbes by binding iron and making it unavailable for microbial development (Stopforth *et al.*, 2005a).

13. INTERVENTIONS DURING PROCESSING

The consumers are looking for foods that are minimally processed, microbiologically safe and chemical free with longer shelf life. In order to meet the consumers' demands, the food researchers and the food industry has developed various natural alternatives (as discussed above) that exhibit strong antioxidant and antimicrobial activity. The use of natural preservatives is not new in the history of food science. The natural food preservatives can be classified into antimicrobial agents and antioxidant agents (Sharif *et al.*, 2017).

14. METHOD OF INCORPORATION OF ANTIMICROBIAL AND ANTIOXIDANT AGENTS

The natural antimicrobials and antioxidants can be used alone or in combination with other natural preservatives. Their efficacy is enhanced when used in combinations with other preservatives.

14.1 By Direct Incorporation into Products

The natural preservatives can be directly incorporated into the meat and other food formulations. In this way, the quantity of active ingredients that is added to the products can be controlled.

14.2 Spraying on the Food Surface

The natural preservatives in solution or the active ingredients can be sprayed on the food surface.

14.3 By Incorporating in Washing Solution

The natural preservatives may be incorporated in the solutions that are used for washing/cleaning of the food.

14.4 By Incorporation into Packaging Materials/Active Packaging

Edible films and coatings often incorporated with active ingredients of natural preservatives for fresh as well as processed products are gaining much interest in the recent era. They are environmental friendly and at the same time provide opportunity for enhancing quality, safety, stability and mechanical handling properties. Apart from the protection it offers to the food, most natural antimicrobial agents and edible films are biodegradable and degrade readily in the environment. An antimicrobial packaging system can be obtained directly by incurporating microbial agents into the packaging film or as coatings of packaging film. The natural antimicrobial that may be incorporated into the antimicrobial packaging can be catagorized as biologically derived components such as bacteriocins, enzymes and plant extracts which are discussed earlier. Nisin and alginate coated film has been shown to reduce *Listeria monocytogenes* in ready to eat turkey products (Juck *et al.*, 2010).

14.5 Regulation of Natural Antimicrobials for Food Application

Most food products require protection against microbial spoilage during storage and consumers also demand safe natural products. This drives the search of food authorities and researchers for mild preservation techniques to improve microbial quality and safety without causing nutritional and organoleptic losses, along with the minimal use of synthetic antibiotics. In this context natural compounds are gaining a great interest from research and industry, due to the potential to provide quality and safety benefits, with a reduced impact on the human health. The numerous experimental applications of essential oils, enzymes, bacteriocins, chitosans, and organic acids to various fresh perishable foods demonstrate that they are well suited to be utilized as preservatives in foods and could be often valid alternatives to synthetic food additives.

The growing menace of antimicrobial resistance amongst the food-borne pathogens is escalating efforts to rely more on these natural agents. However, regulation of the purified form is always a tough question to face. First of all it is necessary to have a good understanding of the mechanism by which antibacterial agents operate. For many natural compounds these information is still lacking. Better understanding of the modes by which antimicrobials can control microorganisms should provide solid grounds for engineering new and upgraded derivatives with optimized potency and stability. Further research is still necessary for specific case study because it is well demonstrated that the combination of more than one active agent not always amplifies the antimicrobial effects. Also, the regulatory agencies have a major role to play for ensuring food safety standards e.g. The United States Food and Drug Administration (USFDA) and China Food Additives Association (CFAA), etc. Countering antimicrobial resistance is a challenge and it can be well targeted with coherent efforts like use of natural agents, provided safety of human life is not at stake.

15. CONCLUSIONS

The food industry is looking for natural preservatives that would replace the already in use synthetic preservatives. It is necessary that to kill or reduce the microbial population in the animal products, as they not only can be a potential risk for food borne infections and intoxications and premature spoilage of food, but because they may pose a threat with respect to the antimicrobial resistance bacteria that may originate at the farm level or incorporated during the processing operations. Scientific investigations and research is going on and it is the need of the hour to lay down guidelines and standards for proper use of these so called natural antimicrobials and antioxidants. More research input is required regarding the safety of these natural preservatives as well and selection of the right kind of natural preservatives so that the organoleptic properties of the food concerned is not compromised. Moreover, the use of antibiotics as growth promoter in animals to improve the meat yield can be achieved with the use of natural compounds in the feed system.

REFERENCES

[1] Adje, E.Y., Balti, R., Guillochon, D. and Nedjar-Arroume, N. (2011). α 67–106 of bovine hemoglobin: A new family of antimicrobial and angiotensin I-converting enzyme inhibitory peptides. *European Food Research and Technology*, 232(4), 637–646.

[2] Aziz, M. and Karboune, S. (2018). Natural antimicrobial/antioxidant agents in meat and poultry products as well as fruits and vegetables: A review. *Critical Rev Food Sci Nutr.*; 58(3):486–511. doi: 10.1080/10408398.2016.1194256.

[3] Céline Verheust, Katia Pauwels, Jacques Mahillon, Donald R. Helinski and Philippe Herman (2010). Contained Use of Bacteriophages: Risk Assessment and Biosafety Recommendations. *Appl. Biosafety*, 15(1), 327.

[4] Chang, Q., Wang, W., Gili, R.Y., Lipstich, M. and Hanage, W.P. (2014). Antibiotics in agriculture and the risk to human health: How worried should we be? Evolutionary Applications. August 2.

[5] Colomer-Lluch, M., Jofre, J. and Muniesa, M. (2011). Antibiotic Resistance Genes in the Bacteriophage DNA Fraction of Environmental Samples. *PLoS ONE 6(3):* e17549. https://doi.org/10.1371/journal.pone.0017549.

[6] Cosgrove, S.E., Sakoulas, G., Perencevich, E.N., Schwaber, M.J., Karchmer, A.W. and Carmeli, Y. (2003). Comparison of mortality associated with methicillin-resistant and methicillin-susceptible Staphylococcus aureus bacteremia: A meta-analysis. *Clin Infect Dis,* 2003, 36(1):53–59.

[7] De Roever, C. (1998). Microbiological safety evaluations and recommendations on fresh produce. *Food Control,* 9(6), 321–347.

[8] Ebrahimi, M. and Darani, K.K. (2013). Essential Oil as natural food preservatives: Antimicrobial and antioxidant applications. Transworld Research Network. Kerala.

[9] Eevuri, T.R. and Putturu, R. (2013). Use of certain herbal preparation in broiler feeds—A review. *Veterinary World* 6: 172–79.

[10] Erginkaya, Z. and Konuray, G. (2017). Natural Preservatives: An Alternative for Chemical Preservative Used in Foods. World Academy of Science, Engineering and Technology, *International Journal of Biological, Biomolecular, Agricultural, Food and Biotechnological Engineering,* 11(4), 311–316.

[11] Founou, L.L., Founou, R.C. and Essack, S.Y. (2016). Antibiotic resistance in the food chain: a developing country-perspective. *Front Microbiol,* 7, 1–19.

[12] Greer, G.G. (2005). Bacteriophage control of foodborne bacteria. *Journal of Food Protection,* 68(5), pp. 1102–1111.

[13] Ibrahim, E.H., Sherman, G., Ward, S., Fraser, V.J. and Kollef, M.H. (2000). The influence of inadequate antimicrobial treatment of bloodstream infections on patient outcomes in the ICU setting. *Chest,* 118(1), 146–155.

[14] Juck, G., Neetoo, H. and Chen, H. (2010). Application of an active alginate coating to control the growth of Listeria monocytogenes on poached and deli turkey products. *International Journal of Food Microbiology,* 142(3), 302-308.

[15] Kang, C.I., Kim, S.H., Kim, H.B., Park, S.W., Choe, Y.J., Oh, M.D., Kim, E.C. and Choe, K.W. (2003). Pseudomonas aeruginosa bacteremia: risk factors for mortality and influence of delayed receipt of effective antimicrobial therapy on clinical outcome. *Clinical Infectious Diseases,* 37(6), pp.745-751.

[16] Kropinski, A.M., Lingohr, E.J., Moyles, D.M., Ojha, S., Mazzocco, A., She, Y.-M., Bach, S.J., Rozema, E.A., Stanford, K., McAllister, T.A. and Johnson, R.P. (2012). Endemic bacteriophages: a cautionary tale for evaluation of bacteriophage therapy and other interventions for infection control in animals. *Virol.* J., 9: 207.

[17] Lafarga, T., Rai, D. K., O'Connor, P. and Hayes, M. (2015). A bovine fibrinogen-enriched fraction as a source of peptides with in vitro renin and angiotensin-i-converting enzyme inhibitory activities. *Journal of Agricultural and Food Chemistry,* 63(39), 8676–8684.

[18] Lawrence D. Goodridge and Bledar Bisha (2011). Phage-based biocontrol strategies to reduce foodborne pathogens in foods. *Bacteriophage,* 1(3): 130 137.

[19] Lone Gram, Lars Ravn, Maria Rasch, Jesper Bartholin Bruhn , Allan B. Christensen and Michael Givskov (2002). Food spoilage-interactions between food spoilage bacteria. *International Journal of Food Microbiology* 78 79–97.

[20] Lorraine Endersen, Jim O Mahony, Colin Hill, R. Paul Ross, Olivia McAuliffe and A Coffey (2014). Phage therapy in the food industry. *Ann. Rev. Food Sci. Technol.,* 5: 327 349.

[21] Luria Leslie Founou, Raspail Carrel Founou and Sabiha Yusuf Essack (2016). Antibiotic Resistance in the Food Chain: A Developing Country-Perspective. *Front Microbiol.*; 7: 1881. doi: 10.3389/fmicb.2016.01881.

[22] McDermott, P.F.1, Zhao, S., Wagner, D.D., Simjee, S., Walker, R.D. and White, D.G. (2002). The food safety perspective of antibiotic resistance. *Anim Biotechnol,* 13(1):71–84.

[23] Mine, Y., Ma, F. and Lauriau, S. (2004). Antimicrobial peptides released by enzymatic hydrolysis of hen egg white lysozyme. *Journal of Agricultural and Food Chemistry,* 52(5), 1088–1094.

[24] Pellegrini, A., Thomas, U., Wild, P., Schraner, E. and Von Fellenberg, R. (2000). Effect of lysozyme or modified lysozyme fragments on DNA and RNA synthesis and membrane permeability of Escherichia coli. *Microbiological Research,* 155(2), 69–77.

[25] Peric, L., Zikic, D. and Lukic, M. (2009). Application of alternative growth promoters in broiler production. *Biotechnology of Animal Husbandry,* 25: 387–97.

[26] Roberts, R.R., Hota, B., Ahmad, I., Scott, R.D., Foster, S.D. and Abbasi, F. (2009). Hospital and societal costs of antimicrobial resistant infections in a Chicago teaching hospital: implications for antibiotic stewardship. *Clin Infect Dis,* 2009, 49(8):1175–1184. doi:10.1086/605630.

[27] Shaltout, F.A., Thabet, M.G. and Hanan, A. (2017). Impact of some essential oils on the quality aspect and shelf life of meat. *J. Nutr Food Sci,* 7(2), 647.

[28] Sharif, N.M., Sreedevi, B., Chaitanya, R.K. and Sreenivasulu, D. (2017). Beta-lactamase antimicrobial resistance in Klebsiella and Enterobacter species isolated from healthy and diarrheic dogs in Andhra Pradesh, India *Vet World.*; 10(8): 950–954.

[29] Sharif, Z.I.M., Mustapha, F.A., Jai, J., Yusof, N.M. and Zaki, N.A.M. (2017). Review on methods for preservation and natural preservatives for extending the food longevity. *Chemical Engineering Research Bulletin,* 19:145–153.

[30] Shen, X., Sun, X., Xie, Q., Liu, H., Zhao, Y., Pan, Y., Hwang, C.A. and Wu, V.C. (2014). Antimicrobial effect of blueberry (*Vaccinium corymbosum* L.) extracts against the growth of *Listeria monocytogenes* and *Salmonella enteritidis*. *Food Control,* 35(1), 159–165.

[31] Singh, P., Sahoo, J., Chatli, M.K. and Biswas, A.K. (2014). Shelf life evaluation of raw chicken meat emulsion incorporated with clove powder, ginger and garlic paste as natural preservatives at refrigerated storage (4±1°C). *International Food Research Journal,* 21(4) 1363–1373.

[32] Smith, R.D. and Coast, J. (2013). The economic burden of antimicrobial resistance: Why it is more serious than current studies suggest. London, London School of Hygiene & Tropical Drug (http://www.lshtm.ac.uk/php/intrafacultyinitiatives/economics/assets/dhamr2012appendix.pdf).

[33] Stopforth, J.D., Skandamis, P.N., Davidson, P.M. and Sofos, J.N. (2005a). Naturally ocurring compounds: Animal sources. In Antimicrobials in Food. Ch. 15, Edited by: Davidson, P.M., Sofos, J.N. and Branen, A.L. 453–505. Boca Raton, FL: CRC Press.

[34] Teng-Hern, A.L. Kok-Gan, C. and Han, L.L. (2014). Application of bacteriophage in biocontrol of major foodborne bacterial pathogens. *J. Mol. Biol. Mol. Imag.*, 1(1): 9.

[35] Thammasirirak, S., Pukcothanung, Y., Preecharram, S., Daduang, S., Patramanon, R., Fukamizo, T. and Araki, T. (2010). Antimicrobial peptides derived from goose egg white lysozyme. Comparative Biochemistry and Physiology Part C: *Toxicology & Pharmacology*, 151(1), 84–91.

[36] Unalan, I.U., Korel, F. and Yemenicoglu, A. (2011). Active packaging of ground beef patties by edible zein films incorporated with partially purified lysozyme and Na2EDTA. *Int J Food Sci Technol*, 46:1289–1295.

[37] Verraes, C., Van Boxstael, S., Van Meervenne, E., Van Coillie, E., Butaye, P., Catry and B., *et al.* (2013). Antimicrobial resistance in the food chain: A review. *International Journal of Environmental Research and Public Health*, 10, 2643e2669.

[38] Watkins, R.R. and Bonomo, R.A. (2016). Overview: Global and Local Impact of Antibiotic Resistance. *Infect Dis Clin North Am.* 30(2):313-322. doi: 10.1016/j.idc.2016.02.001.

[39] World Health Organization (2014). Antimicrobial resistance: global report on surveillance. ISBN 978 92 4 156474 8. http://apps.who.int/iris/bitstream/handle/10665/ 112642/9789241564748_eng.

Exploring the Inhibition of Vitamin B6 Biosynthesis in *Plasmodium falciparum*

Mamta Yadav[1]*, Rahul B. Aher[2], Kunal Roy[2] and Dinkar Sahal[1]

[1]Malaria Drug Discovery Laboratory, International Centre for Genetic Engineering and Biotechnology, Aruna Asaf Ali Marg, New Delhi, India

[2]Drug Theoretics and Cheminformatics Laboratory, Department of Pharmaceutical Technology, Jadavpur University, Kolkata–700 032, India

E-mail: *dinkar@icgeb.res.in

ABSTRACT: The emergence of drug-resistant P. falciparum strains that have rendered first-line drugs ineffective emphasizes the urgent need to search for new lead compounds with novel targets for future anti-malarial drug development. In the quest of novel antimalarials, several libraries of compounds were docked in silico against P. falciparum pyridoxal synthase (PLP synthase) using Schrödinger GLIDE software. The results showed 19 compounds with strong binding affinity with docking score ranging from –9.00 to –11.5. To validate the docking study, selected compounds will be taken for in-vitro enzyme inhibition assay and antiplasmodial activity against CQ sensitive (Pf3D7) and resistant (PfINDO) strains of P. falciparum. Moreover, cytotoxicity of active compounds will be performed against HEK293 and Hela cells lines. The oxidative stress response of selected compounds (SI > 10) will be measured using H_2DCF-DA (2', 7'-dichlorofluorescein diacetate) fluorescent probe. The curative, suppressive and prophylactic anti-malarial efficacy of lead compounds will be evaluated in Plasmodium berghei (ANKA) infected mouse model. We anticipate at the end of the study to identify a lead antimalarial compound that can be further processed for antimalarial drug development.

Keywords: Drug-resistant P. falciparum,Vitamin B6 biosynthesis, PLP synthase, Schrödinger GLIDE, Antiplasmodial, Antimalarial study.

1. INTRODUCTION

Malaria is a life-threatening disease caused by parasites from *Plasmodium* genus, transmitted to human through a mosquito bite. There are five *Plasmodium* species causing malaria in humans, includeing *P. falciparum* responsible for the most reported cases worldwide.[1] According to the World Malaria Report (2016), the African region accounted for 90% of global cases with 92% of deaths. Several approaches are currently used to control this infection. These measures include the use of long-lasting insecticide-treated bed nets for prevention and chemotherapy for treatment. However, since 2009 mounting evidence of the development of *Plasmodium falciparum* resistant strains to the artemisinin-based combination therapies (ACTs), has been observed,[2] highlighting the need for a continuous search for new antiplasmodial leads with a novel mechanism of action for the development of new chemotherapeutics.

In the quest of novel antimalarials, attention has been drawn on selective interference with the parasite's metabolism without harming the human host.[3] In this respect, the biosynthesis of parasite-specific vitamin pathways represents promising drug target.[3,4] Vitamins are molecules with several functions including antioxidants (e.g., ascorbic acid), precursors in electron carrying processes, and acting as cofactors in enzymatic reactions.[5] Mammals generally depend on the uptake of vitamins from their diet, whereas *Plasmodium* synthesizes them *de novo.*[6] In fact, *Plasmodium* possesses three vitamin biosynthetic pathways including vitamin B1 (thiamine), vitamin B6 (pyridoxine) and vitamin B9 (folate).[4,7] Nowadays, several inhibitors of folate metabolism have been reported,[8,9] however, only a few studies have identified potential inhibitors of vitamin B6 biosynthesis pathway.[10]

The metabolically active form of vitamin B6 knows as Pyridoxal-5'-phosphate (PLP) is used as a cofactor by enzymes catalyzing the amino acid decarboxylation, elimination, interconversion, and transamination.[11] In fact, PLP is involved in the protection of biomolecules like lipids, nucleic acids and proteins by itself quenching ROS like the singlet oxygen[12] and the superoxide generated during hemoglobin digestion.[13,14] In addition, erythrocytes are known to be the major carrier of vitamin B6 in humans[15] which raises the question of *Plasmodium* needs to utilize its energy for PLP generation. Studies have shown that vitamin B6 is bound tightly to serum proteins like albumin and hemoglobin[5] and might not be available to the parasite. Moreover, there are no transporters known in *P. falciparum* till date for vitamin B6 uptake from host.[16] Since PLP plays critical roles in several metabolic pathways of the parasite, the inhibition of its biosynthesis can be considered as a possible target for potent inhibitors.[17–19]

This multifunctional metabolite is produced in malaria parasite from two metabolic pathways. The salvage pathway involving the interconversion of its three vitamers (pyridoxal, pyridoxine, pyridoxamine) and their respective phosphate esters by pyridoxal kinase (PdxK) and a phosphatase[7] (Figure 1). While, the *de novo* pathway catalyzed by B6 biosynthesis enzyme, converts L-glutamine, D-ribose-5-phosphate (R5P) and DL-glyceraldehyde 3-phosphate (G3P) into PLP[10] (Figure 1). Thus, the inhibition of enzymes involved in the synthesis of this key metabolite necessary for the survival of the malaria parasite will lead to the increase of ROS responsible for oxidative stress that will be detrimental to *Plasmodium*. Therefore, this study is designed to screen and identify potent inhibitors for the enzymes involved in vitamin B6 biosynthesis.

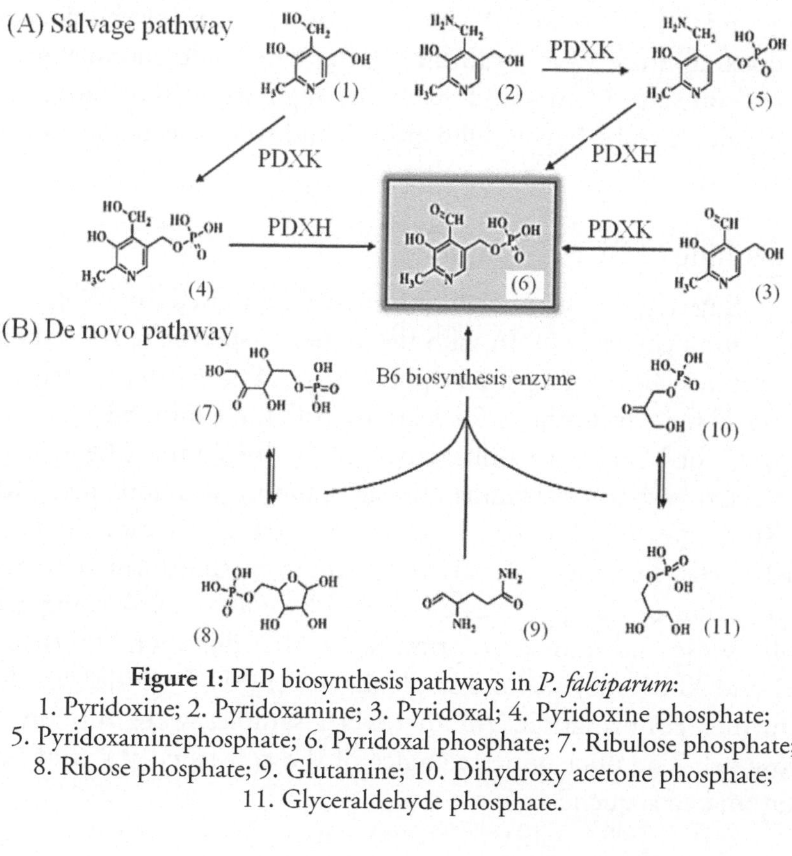

Figure 1: PLP biosynthesis pathways in *P. falciparum*:
1. Pyridoxine; 2. Pyridoxamine; 3. Pyridoxal; 4. Pyridoxine phosphate; 5. Pyridoxaminephosphate; 6. Pyridoxal phosphate; 7. Ribulose phosphate; 8. Ribose phosphate; 9. Glutamine; 10. Dihydroxy acetone phosphate; 11. Glyceraldehyde phosphate.

2. MATERIALS AND METHODS

2.1 In Silico Ligand Screening

A database of commercially available compounds [PubChem and PubMed (217), MMV (400), Supernatural (98), Maybridge (3,018), Drug bank (2,094)], ligand like hits (70) fromiview server, known antiplasmodial and antimalarial compounds (26), compounds present in our laboratory library [Chalcones (77), triazoles (55), indoles (14), quinolines (15), stilbenes (28), chalcone-stilbene-hybrids (22), aplysinopsins (19), imidazoles (23)] and the well-known inhibitors of pyridoxal biosynthesis in yeast and bacteria (10) was created. The *in silico* docking of the molecules in the database was done against the crystal structure of *Pf* Pyridoxal Synthaseusing Schrödinger GLIDE software. Compounds were prioritized by predicting the ADME properties using QikProp module of Maestro software version 11.1 (Schrödinger) taking the *Lipinski rule of five* property as a filter. Compounds, which passed this property, were taken for the docking. The docking study was performed using the Glide module of the Maestro software. Compounds with the high score in negative scale were predicted to be having high affinity and were selected for further experiments.

2.2 Cloning of PLP Synthase

The coding regions of the enzyme were amplified by PCR from *P. falciparum* genomic DNA with the sequence-specific sense and anti-sense primers using Taq DNA polymerase. The cycling parameters were 95°C for 5 min, then 37 cycles of 95°C for 1 min, 53°C for 45 s and 72°C for 1 min with a final step at 72°C for 7 min. The gene was constructed with a 3' terminus His-tag. Cloning was done into Nde1 and Xho1 sites of pET28a (+) vector and transformation in *E. coli* TOP10 cells. Plasmid isolation was done from transformed colonies and insertion of the gene was confirmed by colony PCR using gene-specific sense and anti-sense primers. Double digestion (DD) using Nde1 and Xho1 enzymes and single digestion (SD) by Nde1was done for further confirmation. All the DNA samples were run on 1% agarose gel. The nucleotide sequence of the construct was confirmed by automated sequencing.

2.3 Culturing of P. falciparum

P. falciparum parasite strains (3D7 and INDO) were grown *in vitro* using the method developed by Trager and Jensen[20] with minor modifications. Cultures were maintained in fresh O^+ human erythrocytes at 4% haematocrit at 37°C under reduced O_2 (gas mixture 5% O_2, 5% CO_2, and 90% N_2). cRPMI medium containing 25 mM HEPES supplemented with 0.2% sodium bicarbonate, 0.05% hypoxanthine and 5% albumax I solution was used for culture cultivation. Percent parasitemia was calculated using Giemsa-stained, methanol-fixed thin culture smears. The medium was changed every 24 hrs and the parasitemia was maintained at 8–10%. Cultures at high % parasitemia (>10%) were diluted to low percentage using fresh RBCs.

3. RESULTS AND DISCUSSIONS

3.1 In Silico Docking

To search inhibitors for vitamin B6 biosynthesis enzyme, the compound library was docked onto its crystal structure. After docking, 19 compounds were identified with docking score ranging from –9.00 to –11.5 (Table 1) which were selected for further *in vitro* anti-plasmodial experiments. Supernatural database gave the maximum hits of ~14% with the highest negative docking score of –11.503.

Table 1: Structure of the Selected Compounds with Docking Scores

S. No.	*Name of Compound*	*Structure*	*Origin (Database)*	*Docking Score*
1.	(5-hydroxy-4, 6-dimethylpyridin-3-yl)methyl dihydrogen phosphate	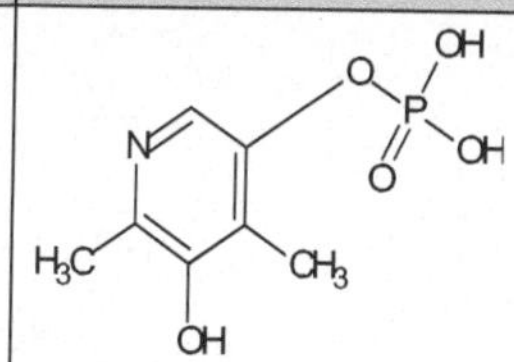	Supernatural	–11.503
2.	2, 3-Diphosphoglycerate	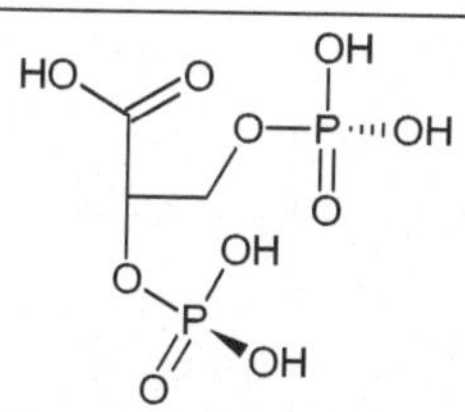	Supernatural	–11.362

(Table contd...)

(Table 1 contd...)

S. No.	Name of Compound	Structure	Origin (Database)	Docking Score
3.	1, 3-Bisphosphoglyceric acid	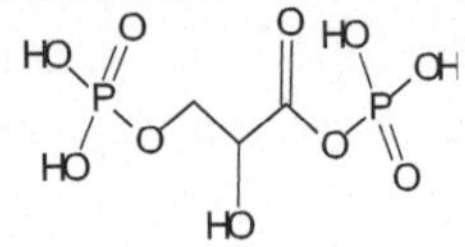	Supernatural	–11.214
4.	(2R, 3R)-2, 4-Dihydroxy-3-phosphonooxybutanoic acid	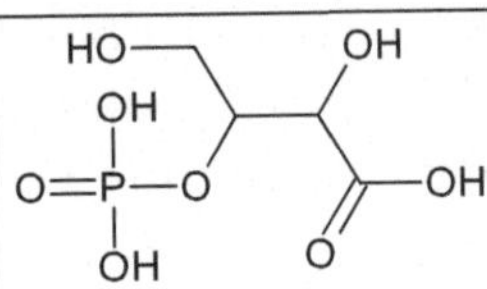	Supernatural	–11.052
5.	3-Phosphoglyceric acid	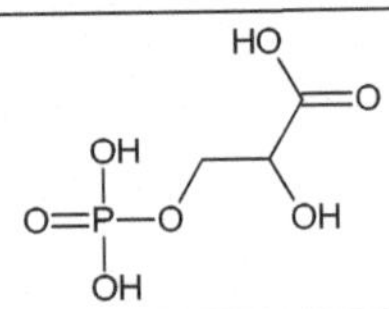	Supernatural	–10.496
6.	Erythrulose 1-phosphate	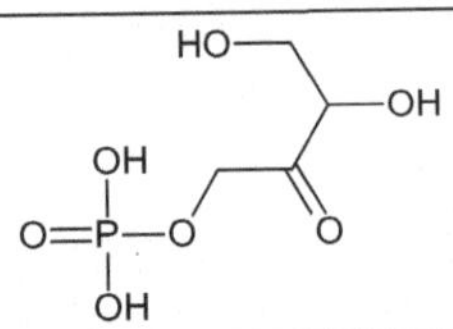	Supernatural	–10.367
7.	3, 4, 5-trihydroxy-2-oxopentyl dihydrogen phosphate	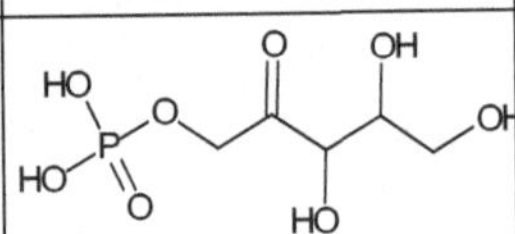	Supernatural	–10.162
8.	2-hydroxy-3-(phosphonooxy) but-3-enoic acid	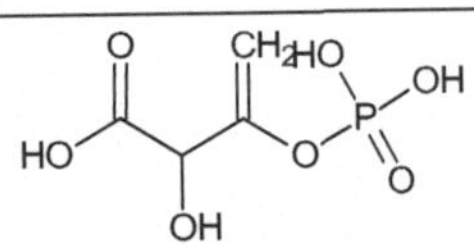	Supernatural	–10.15
9.	Pyridoxal Phosphate Hydrate	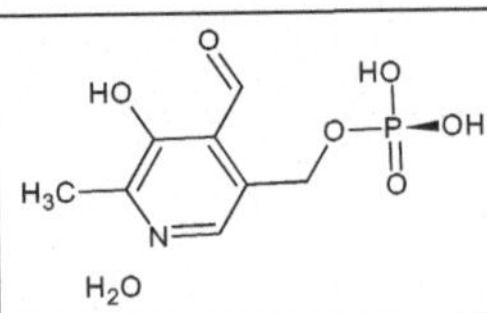	Supernatural	–10.139
10.	2-phosphoglyceric Acid	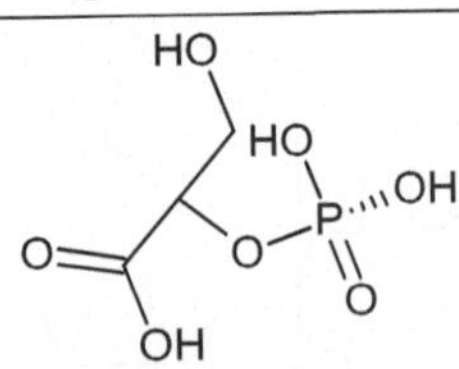	Supernatural	–9.93
11.	D-erythrulose 4-phosphate	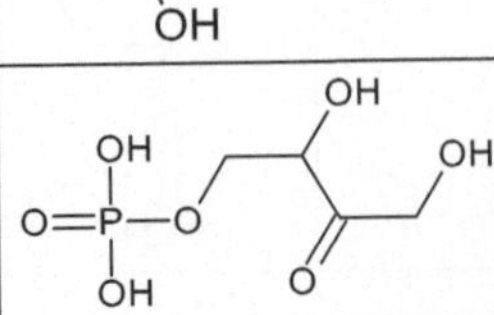	Supernatural and Known inhibitors	–9.682

(Table contd...)

(Table 1 contd...)

S. No.	*Name of Compound*	*Structure*	*Origin (Database)*	*Docking Score*
12.	5-o-phosphonopent-2-ulose	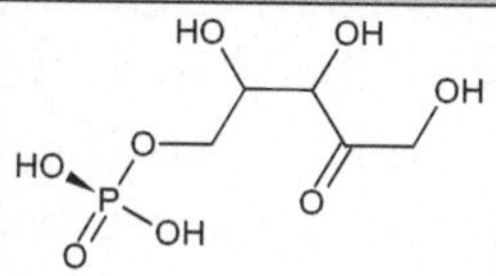	Supernatural	–9.67
13.	3-Phosphohydroxypyruvate	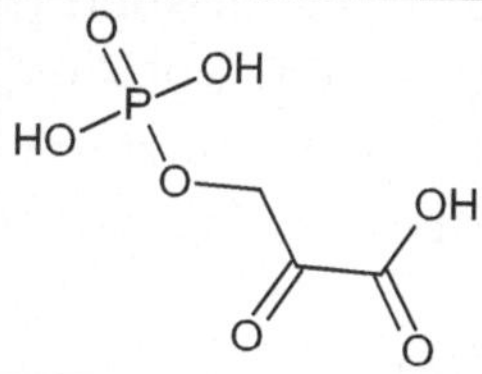	Supernatural	–9.618
14.	3-Glycerophosphate	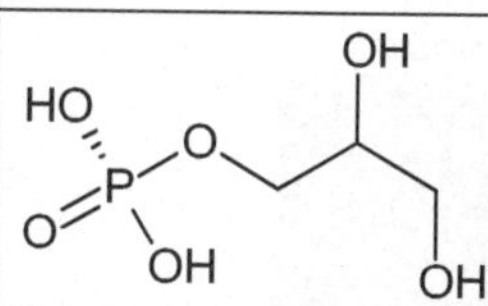	Supernatural	–9.439
15.	2, 3-dioxoindoline-7-carboxylic acid	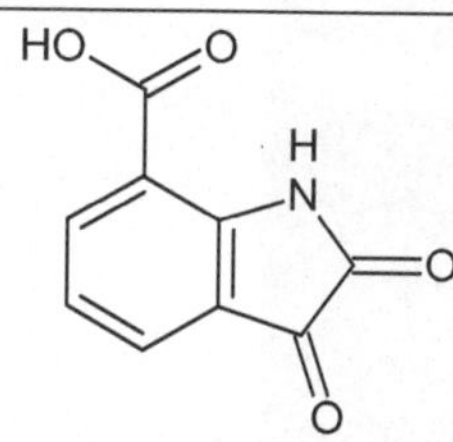	Mabridge	–9.323
16.	6-chloro-1, 2, 3, 4-tetrahydroquinoxaline-2, 3-dione	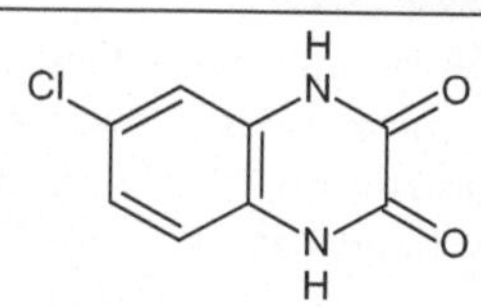	Mabridge	–9.183
17.	4-Deoxypyridoxine 5'-phosphate	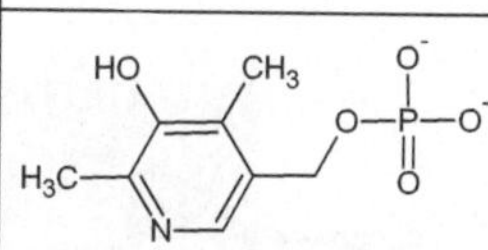	PubChem	–9.13
18.	(6R)-6-amino-6, 8-dihydro-5H-imidazo[1, 2-a]pyrimidin-7-one	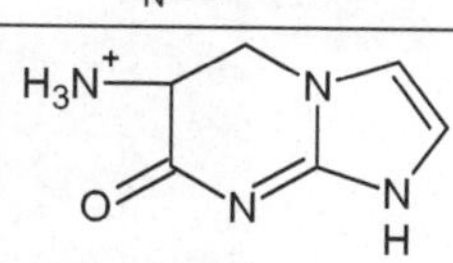	iview	–9.011
19.	1H-Indole-2, 3-dione	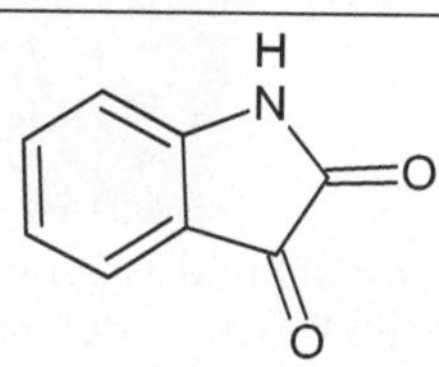	Drug bank	–9.58

3.2 Cloning of Gene

Vitamin B6 biosynthesis enzyme construct of 936 bp was amplified from *Pf* genomic DNA, cloned in pET28a (+) vector of 5.3 kb and transformed in *E. coli* TOP10 cells (Figure 2). Colony PCR of transformed colonies showed the intense band at ~1 kb confirming the presence of the gene in colony plasmid (Figure 3). In comparison with the gel position of single digested plasmid (6.3 kb), the double digested plasmid showed both a 1 kb fragment and a shift of 1 kb in the digested plasmid (5.3 kb) confirming the cloning of gene in vector (Figure 4).

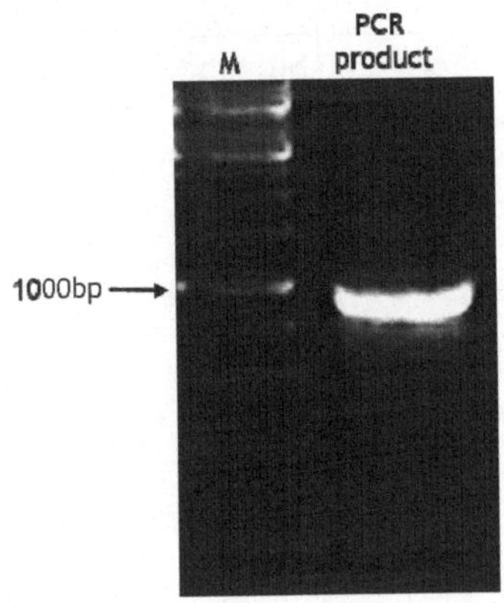

Figure 2: *Pf* genomic DNA was amplified using vitamin B6 biosynthesis enzyme-specific primers and run on 1% agarose gel giving intensified band at ~1 kb.

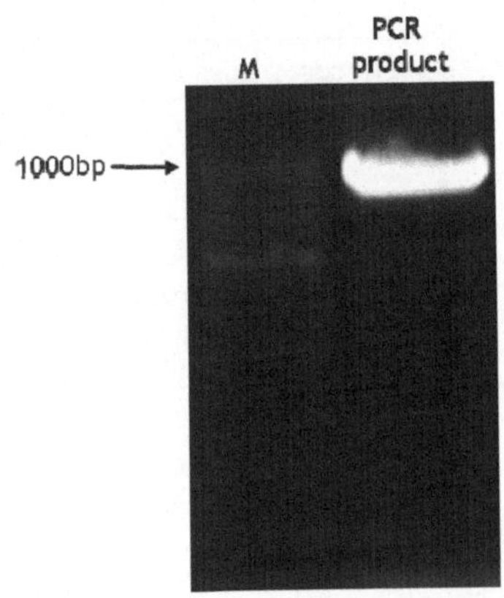

Figure 3: Transformed colony plasmid was amplified using gene-specific primers and run on 1% agarose gel giving intensified band at ~1 kb confirming the presence of the gene in the colony.

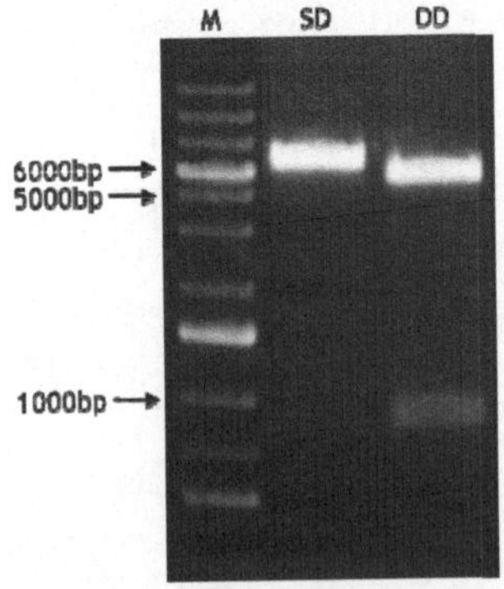

Figure 4: Transformed colony plasmid was isolated and single digested (SD) with Nde1 and double digested (DD) with Nde1 and Xho1. SD giving band at ~1 kb higher position than DD confirming the presence of 1 kb gene in plasmid. DD giving band fallout at ~1 kb specific to cloned gene.

4. CONCLUSION

This study was designed to investigate the inhibitors PLP synthase, the main enzyme of the *de novo* biosynthesis vitamin B6. Nineteen (19) hit compounds with good docking scores including 2, 3-diphosphoglycerate and erythrulose 1-phosphate were identified from 10 different databases. The analogs of these compounds could be taken for *in vitro* enzyme inhibition assay followed by antiplasmodial and antimalarial activities. At the end of this study, identification of one or more lead antimalarial compounds that can further be processed for drug development can be anticipated.

5. AUTHOR CONTRIBUTION

Dinkar Sahal and Mamta Yadav designed the study; Mamta Yadav identified, prepared libraries of compounds for docking, analyzed the data and prepared the manuscript. Rahul B Aher and Kunal Roy performed the *in silico* docking analysis. Dinkar Sahal provided scientific inputs.

ACKNOWLEDGMENTS

We acknowledge DBT (Department of Biotechnology, India) for providing the scholarship and ICGEB (International Centre for Genetic Engineering and Biotechnology, India) for providing the core-funds for this study.

REFERENCES

[1] Müller, I.B., Knöckel, J., Groves, M.R., *et al.* The assembly of the plasmodial PLP synthase complex follows a defined course. *PLoS One.* 2008; 3(3). doi:10.1371/journal.pone.0001815.

[2] Dondorp, A.M., Nosten, F., Yi, P., *et al.* Artemisinin Resistance in Plasmodium falciparum Malaria. *N. Engl J. Med.* 2012; 361(5): 455-467. doi:10.1056/NEJMoa0808859.Artemisinin.

[3] Müller, I.B., Wu, F., Bergmann, B., *et al.* Poisoning pyridoxal 5-phosphate-dependent enzymes: A new strategy to target the malaria parasite Plasmodium falciparum. *PLoS One.* 2009;4(2). doi:10.1371/journal.pone.0004406.

[4] Wrenger, C., Knöckel, J., Walter, R.D. and Müller, I.B. Vitamin B1 and B6 in the malaria parasite: Requisite or dispensable? *Brazilian J. Med Biol Res.*, 2008; 41(2):82–88. doi:10.1590/S0100-879X2008005000006.

[5] Friedrich W. *Vitamins*. De Gruyter; Berlin, New York, 1988. https://books.google.co.in/books/about/Vitamins.html?id=VS9YWqnMpNsC&redir_esc=y. Accessed July 11, 2017.

[6] Gengenbacher, M., Fitzpatrick, T.B., Raschle, T., *et al.* Vitamin B6 biosynthesis by the malaria parasite Plasmodium falciparum: Biochemical and structural insights. *J. Biol Chem.* 2006; 281(6):3633-3641. doi: 10.1074/jbc.M508696200.

[7] Kronenberger, T., Lindner, J., Meissner, K.A., *et al.* Vitamin B6-dependent enzymes in the human malaria parasite plasmodium falciparum: A druggable target? *Biomed Res Int.* 2014; 2014 (January). doi:10.1155/2014/108516.

[8] Hyde, J.E. Exploring the folate pathway in Plasmodium falciparum. *Acta Trop.* 2005; 94(3 SPEC. ISS.):191-206. doi:10.1016/j.actatropica.2005.04.002.

[9] Nzila, A., Ward, S.A., Marsh, K., Sims, P.F.G. and Hyde, J.E. Comparative folate metabolism in humans and malaria parasites (part I): Pointers for malaria treatment from cancer chemotherapy. *Trends Parasitol,* 2005; 21(6): 292–298. doi:10.1016/j.pt.2005.04.002.

[10] Reeksting, S.B., Uller, I.B., Burger, P.B., *et al.* Exploring inhibition of Pdx1, a component of the PLP synthase complex of the human malaria parasite Plasmodium falciparum. *Biochem J.* 2013; 449:175–187. doi:10.1042/BJ20120925.

[11] Percudani, R. and Peracchi, A. A genomic overview of pyridoxal-phosphate-dependent enzymes. *EMBO Rep.* 2003; 4(9):850-854. doi:10.1038/sj.embor.embor914.

[12] Knöckel, J., Müller, I.B., Butzloff, S., Bergmann, B., Walter, R.D. and Wrenger, C. The antioxidative effect of de novo generated vitamin B6 in Plasmodium falciparum validated by protein interference. *Biochem J.*, 2012; 443(2):397–405. doi:10.1042/BJ20111542.

[13] Ehrenshaft, M., Chung, K.R., Jenns, A.E. and Daub, M.E. Functional characterization of SOR1, a gene required for resistance to photosensitizing toxins in the fungus Cercospora nicotianae. *Curr Genet.* 1999; 34(6):478-485. doi:10.1007/s002940050423.

[14] Jain, M., Mathur, G., Koul, S., Sarin, N.B. Ameliorative effects of proline on salt stress-induced lipid peroxidation in cell lines of groundnut (Arachis hypogaea L.). *Plant Cell Rep.* 2001; 20(5):463–468. doi:10.1007/s002990100353.

[15] Fonda, M.L. and Harker, C.W. Metabolism of pyridoxine and protein binding of the metabolites in human erythrocytes. *Am J. Clin Nutr.* 1982; 35(6):1391–1399. http://www.ncbi.nlm.nih.gov/pubmed/7081120. Accessed July 11, 2017.

[16] Gardner, M.J., Hall, N., Fung, E., *et al.* Genome sequence of the human malaria parasite Plasmodium falciparum. *Nature,* 2002; 419 (6906): 498–511. doi:10.1038/nature01097.

[17] Burkhard, P., Tai, C.-H., Jansonius, J.N. and Cook, P.F. Identification of an allosteric anion-binding site on O-acetylserine sulfhydrylase: structure of the enzyme with chloride bound. *J. Mol Biol.* 2000; 303(2):279–286. doi:10.1006/jmbi.2000.4109.

[18] Krupka, H.I., Huber, R., Holt, S.C. and Clausen, T. Crystal structure of cystalysin from Treponema denticola: a pyridoxal 5'-phosphate-dependent protein acting as a haemolytic enzyme. *EMBO J.*, 2000; 19(13):3168–3178. doi:10.1093/emboj/19.13.3168.

[19] Storici, P., De Biase, D., Bossa, F., *et al.* Structures of γ-Aminobutyric Acid (GABA) Aminotransferase, a Pyridoxal 52-Phosphate, and [2Fe-2S] Clustercontaining Enzyme, Complexed with γ-Ethynyl-GABA and with the Antiepilepsy Drug Vigabatrin. *J. Biol Chem.* 2004; 279(1):363-373. doi:10.1074/jbc.M305884200.

[20] Trager, W. and Jensen, J.B. Human malaria parasites in continuous culture. *Science (80-).* 1976; 193(4254):673 LP-675. http://science.sciencemag.org/content/193/4254/673.abstract.

ABBREVIATIONS

Pf	-	*Plasmodium falciparum*
HEK	-	human embryonic kidney
PLP	-	pyridoxal-5-phosphate
SI	-	selectivity index
H_2DCF-DA	-	2',7'-dichlorofluorescein diacetate
ACTs	-	artemisinin-based combination therapies
ROS	-	reactive oxygen species
PdxK	-	pyridoxal kinase
R5P	-	D-ribose-5-phosphate
G3P	-	DL-glyceraldehyde 3-phosphate
MMV	-	medicines for malaria venture
ADME	-	absorption, distribution, metabolism and excretion
PCR	-	polymerase chain reaction
DNA	-	deoxyribonucleic acid
SD	-	single digestion
DD	-	double digestion
cRPMI	-	roswell park memorial institute medium
HEPES	-	4-(2-hydroxyethyl)-1-piperazineethanesulfonic acid
RBC	-	red blood cells.

LUDHIANA RESOLUTION

Antimicrobial Resistance and Strategies for its Mitigation

LUDHIANA RESOLUTION
on
Antimicrobial Resistance and Strategies for its Mitigation

We, the Delegates of the 4-days International Training Workshop on 'Antimicrobial Resistance and Strategies for its Mitigation' organised by the Centre for Science & Technology of the Non-Aligned and Other Developing Countries (NAM S&T Centre), New Delhi in partnership with Guru Angad Dev Veterinary & Animal Sciences University (GADVASU), Ludhiana (Punjab), India during 10–13 July 2018.

Comprising the clinicians, microbiologists, veterinarians, pharmacists and experts in allied sciences, public health experts and policy planners working in academic and R&D institutions, government departments and ministries, NGOs and industries of the developing countries from Afghanistan, Egypt, India, Indonesia, Iran, Kenya, Malaysia, Mauritius, Myanmar, Nepal, Nigeria, Palestine, South Africa, Sri Lanka and Zambia.

Expressing Gratitude to the organisers of the International Training Workshop.

Having Considered that the antimicrobial resistance is posing a serious threat to human, animal and environmental health and the economy globally, and the systematic misuse of the antimicrobial drugs in human and veterinary medicine and food production have put every nation at risk, particularly the developing countries. In addition, the availability of falsified and substandard medicines further contributes to the development of antimicrobial resistance.

Recognising that it is essential to provide effective antimicrobial drugs for protecting humans and animals from diseases and ensuring the success of surgery, chemotherapy and other medical interventions.

FURTHER RECOGNISING THAT the antimicrobial resistance is emerging and spreading at a threateningly fast pace through microbial populations and even though the development of resistive strains is a normal evolutionary process, it is getting accelerated by the selective pressure exerted by widespread use of various antimicrobial drugs.

AFTER EXTENSIVELY DELIBERATING ON the urgent need to strengthen the existing strategies and develop innovative and alternative strategies to combat the detrimental effect of antimicrobial resistant micro-organisms under the One Health approach.

COMMITTING TO achieve the development goal set forth to evolve strategies for the mitigation of resistance to antimicrobial agents and foster international, regional, sub regional, multilateral and bilateral collaboration in evolving and implementing the same.

UNANIMOUSLY RESOLVE AND RECOMMEND that:

- Accelerated efforts should be made to comprehensively examine the current status and understand the real extent of antimicrobial resistance in human and animal pathogens and environmental micro-organisms.
- More basic and applied research work should be directed towards evolving newer and effective strategies for prevention and containment of antimicrobial resistance, particularly in pathogenic micro-organisms.
- Ministries and agencies concerned with the human and animal health and environment should constitute scientific groups comprising the experts and practicing professionals of the NAM member countries and other developing countries to address the problems from various angles and also to exchange technological advancements available to mitigate antimicrobial resistance problems.
- Specialised centres and research laboratories exclusively for the identification and surveillance of antimicrobial resistance in humans, animals and environment should be set up in each country or region to facilitate exchanging their antibiotic resistance data on a common platform like GLASS by WHO on a regular basis to enable interventions as well as knowledge sharing and networking.
- Concerted efforts should be made to build strong, sustainable quality systems to eliminate falsified and substandard antimicrobials and ensure the supply of quality medicines, while

strengthening post-market quality surveillance to guarantee their accessibility to the end users.

- Research should be focused on finding new and innovative alternatives, including but not limited to, bacteriophages, antibodies, herbal medicines or newer and safer antibiotics to address the problem of antimicrobial resistance.
- Specialized centres and research laboratories exclusively for the development of mitigation strategies against antimicrobial resistance like novel therapies, prophylactics and diagnostics should be set up by the governments of the different countries for mutual cooperation and networking among the institutions and nations.
- Established specialised centres and research laboratories should work in close tandem with the government agencies to formulate appropriate policies and guidelines to prevent escalation of menace of antimicrobial resistance.
- Emphasis should be laid on devising and enforcing appropriate legislative control measures to avoid the misuse of the antimicrobials in both human and veterinary sectors.
- The implementation of Antimicrobial Stewardship and infection prevention and control programme in human and veterinary medicine should be accelerated to reduce the development of antimicrobial resistance and its spread.
- The Governments and NGOs should be actively involved in the creation of public awareness through knowledge transfer, including but not limited to print and mass media, regarding the judicious use of antimicrobials.

Thus, Resolved and Adopted on this Day, the 13th of July 2018 at Ludhiana (Punjab), India.

Author Index

S

W

Y

9 789389 934168

Printed at Repro India Ltd.